T0234703

SpringerBriefs in Optimization

Series Editors

Sergiy Butenko, Texas A&M University, College Station, TX, USA

Mirjam Dür, University of Trier, Trier, Germany

Panos M. Pardalos, University of Florida, Gainesville, FL, USA

János D. Pintér, Lehigh University, Bethlehem, PA, USA

Stephen M. Robinson, University of Wisconsin-Madison, Madison, WI, USA

Tamás Terlaky, Lehigh University, Bethlehem, PA, USA

My T. Thai ⓘ, University of Florida, Gainesville, FL, USA

SpringerBriefs in Optimization showcases algorithmic and theoretical techniques, case studies, and applications within the broad-based field of optimization. Manuscripts related to the ever-growing applications of optimization in applied mathematics, engineering, medicine, economics, and other applied sciences are encouraged.

More information about this series at http://www.springer.com/series/8918

Maude Josée Blondin

Controller Tuning Optimization Methods for Multi-Constraints and Nonlinear Systems

A Metaheuristic Approach

 Springer

Maude Josée Blondin
Département Génie Électr. & Inform.
Université de Sherbrooke
Sherbrooke, QC, Canada

ISSN 2190-8354   ISSN 2191-575X   (electronic)
SpringerBriefs in Optimization
ISBN 978-3-030-64540-3   ISBN 978-3-030-64541-0   (eBook)
https://doi.org/10.1007/978-3-030-64541-0

Mathematics Subject Classification: 93C05, 93C10, 93C35, 93C73, 93C95, 68T01, 68T40, 68T20, 74P99, 65K10, 34H05

This Springer imprint is published by the registered company Springer Nature Switzerland AG.
The registered company address is: Gewerbestrasse 11, 6330 Cham, Switzerland

Preface

Control theory, that is, controller design and tuning, has always been faced with the challenges of meeting various time and frequency domain requirements to achieve desired behaviors. Similarly, several constraints, some of a physical nature such as the saturation of actuators, while others related to operations such as cost of production, must be observed. Classical control systems and traditional tuning methods are sometimes inadequate to meet all system requirements. As a result, more complex controllers have been designed for nonlinear systems with multi-constraints. It has been demonstrated that optimization approaches based on metaheuristics are promising tuning tools.

This monograph aims to cover in depth the application of metaheuristics to controller tuning and presents future research direction on the topic. The book explains how to pass from the theory to the application of metaheuristics in practice. Explanations and discussions on two practical case studies provide training for researchers in this field. Thus, this monograph can be a textbook used for graduate courses on the topics. The monograph is divided into four chapters, each of which stands on its own.

The first chapter presents classical controller structures along with their classical tuning approaches. The challenges faced in control engineering are exposed, highlighting the need for developing and applying optimization techniques to controller tuning. Besides, this chapter presents the distinction between different categories of optimization techniques to empower the reader to select the most appropriate technique for the problem at hand. An emphasis is placed on metaheuristics and their features. It is also explained step by step how to model controller tuning as an optimization problem.

The second chapter presents a brief history of the evolution of metaheuristics, along with a comprehensive survey of metaheuristics. The main characteristics and principles common to any metaheuristics are presented, providing the core knowledge to understand any metaheuristics pattern. Two of the most popular metaheuristics are also detailed, that is, genetic algorithm (GA) and simulated annealing (SA), respectively, a population-based algorithm and a trajectory-based algorithm.

In the third chapter, it is explained how to apply metaheuristics to controller tuning. The system studied in this monograph are two canonical benchmark problems in control theory that deal with nonlinearities and multi-constraints, that is, the inverted cart pendulum (ICP) and the automatic voltage regulator (AVR) system. SA and GA are employed to tune several controller structures for the ICP and AVR system. This chapter highlights the principal features of metaheuristics for controller tuning. Moreover, a new optimization framework for robust optimization is proposed, and its efficiency is demonstrated on the ICP. It is shown that controller tuning by metaheuristics handles perturbation management and robustness criteria and yields to high dynamic performances. The performances of SA and GA are compared, and the characteristics of each algorithm are underlined. Conjointly, elements for proper algorithm benchmarking are presented. Besides, components that influence metaheuristics performance when applied to controller tuning are analyzed in depth and demonstrated through simulations.

The fourth chapter offers research perspectives and future direction for optimization algorithms applied to controller tuning and metaheuristics in general. New trends in research are also presented, such as hybrid metaheuristics.

This book provides the reader with the knowledge to apply optimization algorithms to controller tuning. It also gives insight into the development of new optimization approaches and the theoretical development needed in this research field.

Sherbrooke, QC, Canada   Maude Josée Blondin
September 2020

Acknowledgments

I would like to thank everyone who has given advice and support to write this monograph. I am also grateful to all the reviewers who took the time to provide feedback to improve this monograph's quality. I wish to thank Dr. Pardalos for his assistance and useful comments throughout the process of this book. Finally, I would like to thank Elizabeth Loew from Springer for her help and support to publish this monograph. This work was in part supported by The Postdoctoral research scholarship from the Fonds de recherche Nature et technologies du Québec.

Contents

Chapter 1
Optimization Algorithms in Control Systems

1.1 Introduction to Control System

In engineering, control refers to systems that control, manage, or regulate systems and processes. Control systems have played an important role in modern technology and society's growth and progress over the past decades. Indeed, they are present in various engineering fields, such as mechanics, aerospace, and electronics. Their spectrum of use is extensive; it goes from controlling simple systems such as an on–off switch for temperature [1] to more complex systems like industrial processes [2] with hundreds of inputs and outputs.

Figure 1.1 presents a closed-loop control system; system's outputs affect the system behavior.

Different structures and methods to design closed-loop control systems exist, but the principle remains the same. The system exist output is compared to the desired response. The difference between both values is the system error, which feeds the control system. The system automatically takes action to bring this error to zero to reach the desired response. Closed-loop control systems are more reliable, accurate, and robust to external disturbances and noise than open-loop control systems. However, closed-loop control systems are more complex to design, usually need more maintenance, and might become unstable.

Closed-loop control systems have enabled the automation of systems. Indeed, closed-loop control systems are the foundation of automatic control. A typical household item that illustrates automatic control is a heater. A sensor continuously monitors the house temperature. A feedback loop, as presented in Fig. 1.1, compares the house temperature to the desired temperature. When the temperature drops below the desired temperature, the heater turns on and stops automatically when the set temperature is reached.

As mentioned above, there exist several types of control structures used in closed-loop systems. The Proportional-Integral-Derivative (PID) controller is one of the

© Springer Nature Switzerland AG 2021
M. J. Blondin, *Controller Tuning Optimization Methods for Multi-Constraints and Nonlinear Systems*, SpringerBriefs in Optimization,
https://doi.org/10.1007/978-3-030-64541-0_1

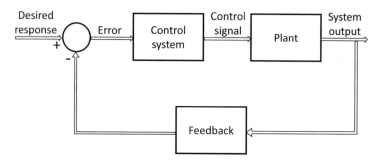

Fig. 1.1 Closed-loop control system

most used in practice due to its ease of use, robustness, and reliability. Applications range from electronic devices to industrial processes [3]. The PID control function is as follows:

$$u(t) = K_p e(t) + K_i \int_0^t e(t)dt + K_d \frac{de(t)}{dt} \tag{1.1}$$

The control signal, $u(t)$, is computed from the system error $e(t)$. The PID controller's performance relies on its proportional gain value K_p, integral gain value K_i, and derivative gain value K_d. Consequently, the tuning of PID controllers is a fundamental step in the control system design. Even though there are only three tunable gains, the tuning is challenging because conflicting criteria related to stability, robustness, and dynamic performance must be satisfied. Therefore, the controller tuning objective is to reach a satisfying trade-off between the system criteria, which applies to PID controllers and also any other types of controllers. Conflicting criteria is one of the reasons several methods for controller tuning exist. Table 1.1 presents some of the so-called classical tuning methods for PID controllers along with practical applications. The second column contains a reference for more details regarding the method, and the last column provides a reference for the method applied to a specific system.

The most known method is the Ziegler–Nichols (ZN), a heuristic that provides rules to set the gains. This method also includes rules to tune P controller, PI controller, and PD controller [21]. These classical methods have been beneficial for control tuning, but they have many drawbacks and limitations. In particular, the ZN method and the other methods presented in Table 1.1 typically respond to particular performance criteria such as the Integral of the Absolute Value of Error. They also hardly handle system constraints and restrictions such as controller saturations, which are present in practice [22, 23]. In the same direction, the classical PID controller also has limitations related to dynamic performance, stability, and robustness. Consequently, more complex PID controllers have been developed to improve control performance, such as two-degree-of-freedom PID (2DOF-PID), fractional order PID (FOPID), and PID with anti-windup. These structures have

Table 1.1 PID controller classical tuning methods and applications

Classical methods	Ref.	System	Ref.
Ziegler–Nichols	[4]	Speed control of Direct Current motor	[5]
Kappa–tau tuning	[6]	Activated sludge aeration process	[7]
Pole placement	[8]	Voltage regulator systems	[9]
Gain and phase based design	[10]	Unstable first-order plus dead-time process	[11]
D-partitioning	[12]	First-order plus dead-time process	[13]
Nyquist based design	[14]	Closed-loop plant	[15]
Cohen–Coon	[16]	First-order plus dead-time process mode	[17]
Internal model control	[18]	Load frequency control of power systems	[19]
Frequency-loop shaping	[20]	Four process models	[20]

extra parameters. For instance, 2DOF-PID has two additional parameters that give control systems greater freedom, enabling the system to achieve faster disturbance rejection without decreasing dynamic performance. However, the classical PID tuning methods are inappropriate for tuning these PID-based controllers. A general approach is to tune the PID gains with a classical approach and, afterward, adjust the extra parameters. However, sequential tuning may provide poor performances [24, 25]. As a result and because more complex control systems have evolved to meet technology needs and their system performances depend on well-tuned parameters, control tuning has been a research topic of interest over the past decades.

1.2   Control Design Challenges

The challenge in control design is twofold: (1) having a controller structure capable of dealing with different requirements such as robustness, dynamic performance, and system uncertainties and (2) having an adequate method to tune the controller structure parameters. As a result, researchers have been devising control structures and crafting optimization algorithms to facilitate control tuning. Figure 1.2 presents the interactions between the development of new control algorithms and tuning algorithms. The system and application requirements drive the creation of control algorithms. Reciprocally, new control algorithms enable new system operations and applications. In the same way, the creation of optimization algorithms supports the development of control methods, facilitating the implementation of novel control systems for new system operations and applications. An example that illustrates all these interactions is the conception of a control system and tuning method for blood glucose levels prediction to control insulin dispensing [26].

Control algorithms can be divided into many categories. Figure 1.3 presents a classification of some of the newest control methods, the so-called modern control methods [27]. These are alternatives to classical PID-based controllers.

Fig. 1.2 Interactions
between the development of
tuning and control algorithms

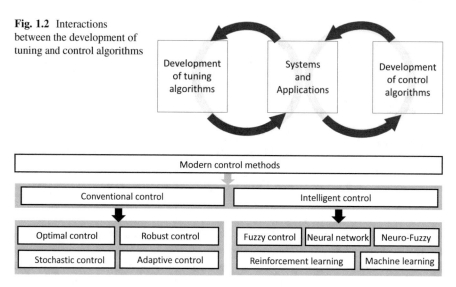

Fig. 1.3 Classification of control theory

There are many ways to categorize control methods. In this monograph, the modern methods are split into two categories: conventional control and intelligent control [28]. Conventional control refers to techniques that require a precise mathematical model of the system to be effective. Therefore, all system dynamics should be modeled mathematically to use a conventional control technique. On the opposite, intelligence control methods do not need an elaborated model. Intelligent control methods model abstractly the system based on system inputs and evaluate the obtained system outputs [28]. For space purposes, Fig. 1.3 presents only some of the most known methods in control theory. The methods could have also been divided once more into linear vs nonlinear control.

1.3 Optimization-Based Approach for Controllers Tuning

Regardless of the control method, all methods have parameters/coefficients to set. For instance, the Linear-Quadratic Regulator (LQR), an optimal control method, minimizes a quadratic function. The quadratic function has two weighting matrices that need to be adequately specified to meet the closed-loop performance specifications [29]. Usually, the LQR designer tries different weights through an iterative process. However, the selection of the weighting matrices may become a hurdle. Similarly, the H-∞ control, a robust control method, also possesses two weighting matrices to set to reach satisfactory closed-loop dynamics [30]. Along the same line, fuzzy control based methods and neural network parameters also require tuning. A promising solution has been using optimization algorithms to tune controller structures to solve this common tuning issue. Indeed, as technology is growing and computers become more and more powerful, hundreds of new optimization algorithms have been proposed to fulfill the needs in many engineering fields, including optimization in control engineering [31]. To employ optimization algorithms, control tuning has to be designed as an optimization problem, consisting of several steps. Figure 1.4 presents the principal steps to design the optimization problem. Selecting the appropriate technique for solving an optimization problem is the key to reach satisfactory results. These steps will help select the most suitable category of optimization techniques to solve the problem.

Indeed, it is essential to determine in which category the optimization problem falls regarding convexity, gradient accessibility, constraints, and the number of objective functions as specific optimization methods exist for each category. For example, there are optimization techniques that are made to handle constraints, while some are not. It is, therefore, essential to identify them, if any. Constrained problems may contain inequality constraints, equality constraints, or both. In a controller tuning context, constraints may include bounds on controller parameter values, control action signals, and phase/gain margins.

The number of objective functions to minimize is also determinant when choosing an optimization technique. There are two popular ways to solve a multi-objective optimization problem. The first approach combines all the objectives into a single objective using weighting, which brings the optimization problem to solving a single-objective function. The weights have to be carefully chosen since they influ-

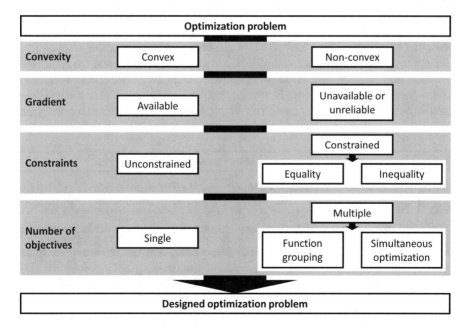

Fig. 1.4 Optimization problem design

ence the optimization search, which subsequently will affect the optimized solution. The second approach optimizes all the objectives simultaneously and chooses the best trade-off between objectives by exploring the Pareto Front, for example. Once the optimization problem is designed, its mathematical representation is as follows:

$$\underset{x \in \Re^n}{\text{minimise}} \quad f_i(x) \qquad (i = 1, 2, \ldots, M)$$

$$\text{subject to} \quad \phi_j(x) = 0 \quad (j = 0, 1, 2, \ldots, J) \qquad (1.2)$$

$$\psi_k(x) \leq 0 \quad (k = 0, 1, 2, \ldots, K)$$

where $f_i(x)$ are the functions to minimize and i refers to the number of functions, and x is the decision vector of n variables. For controller tuning, the functions are the criteria assessing the controllers' performance, and x are the parameters to tune. The designer has to define the objective function. $\phi_j(x)$ and $\psi_k(x)$ are the equality and inequality constraints, respectively, where j and k define the number of constraints. If $j = k = 0$, the optimization problem is unconstrained. Given that the optimization problem is defined, an optimization algorithm must be selected to solve the problem. Several distinctions between optimization algorithms. Figure 1.5 presents a classification of the different types of optimization algorithms exist.

Deterministic algorithms follow a precise procedure, meaning for the same initial conditions, the process/optimization path and the optimized solution will be identical every time the algorithm is run. The deterministic methods can be separated into two categories: the gradient-based methods and the gradient-free methods. Gradient-based methods use gradient information during the optimization process as opposed to gradient-free methods [32]. Contrary to deterministic methods, stochastic algorithms have random attributes. This randomness implies that the optimization process/path will be different at each execution, and the optimized solution may slightly vary, although the initial conditions are identical. Therefore, the same optimization process is not duplicable. Usually, stochastic methods are gradient-free methods. Two categories divide stochastic algorithms: the heuristic

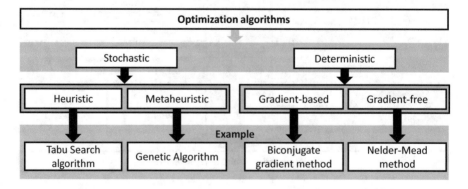

Fig. 1.5 Optimization algorithms classification

methods and the metaheuristics. Heuristics can be defined as techniques designed for reaching a satisfactory solution in reasonable computation time. The optimal solution is not guaranteed, but heuristics are good alternatives for solving problems for which deterministic methods are computationally too burdensome or fail to converge. The design of more complex heuristic strategies has created the category of metaheuristic algorithms. A high level of abstraction characterizes this type of algorithms, which allows them to be easily adapted and applied to a wide range of different problems. They are specially designed to tackle complex and nonlinear optimization problems for which deterministic and heuristics optimization fails to provide satisfactory results [33]. They are known to bypass areas of local minima in the case of multimodal functions effectively. Moreover, metaheuristics are, in general, simple to implement. They can handle constraints as well as convex and non-convex functions. They are usually gradient-free methods and can be used for single-objective optimization as well as for multi-objective optimization. While convergence to the optimal solution is not guaranteed, metaheuristics reach satisfactory solutions, usually near the optimal, in reasonable computation time.

For these reasons, metaheuristics have attracted much attention in the past two decades. Researchers have applied them to a wide range of domains, including medical, transportation, and image processing [34]. Metaheuristics have positively impacted the field of control engineering. Indeed, metaheuristics are suitable to tune control systems designed from classical control methods to modern control methods. Therefore, this monograph aims to present the latest development in control tuning performed by metaheuristics and be a reference textbook to apply metaheuristics to controller tuning. It provides the reader with the core knowledge concerning the most important metaheuristics features in control engineering. The future research direction that the community should take to bring the metaheuristic optimization to control engineering to a mature scientific level is also presented. The remaining chapters of this monograph are organized as follows:

Chapter 2 introduces the most important principles behind any metaheuristics. The evolution and origin of metaheuristics, along with their source of inspiration, are explained. An exhaustive list of metaheuristics is presented, and their first application is surveyed. A general metaheuristics framework is presented along with explanations for control system tuning. Two of the most known metaheuristics, Genetic Algorithm (GA) and Simulated Annealing (SA), are presented. Chapter 3 contains two case studies, i.e., the Automatic Voltage Regulator (AVR) system and the Inverted Cart Pendulum (ICP), for which GA and SA tune their control systems. The AVR and ICP are canonical and fundamental benchmark problems in control theory, making them excellent systems to highlight the strength of metaheuristics tuning. Metaheuristics components and their influential factors are studied and analyzed. Chapter 4 presents the future direction and research gaps in the field of metaheuristics in general with attention on hybrid algorithms for control tuning.

References

1. Golob, M., Tovornik, B., Donlagic, D.: Comparison of the self-tuning on-off controller with the conventional switching controllers. In: Proceedings 1992 The First IEEE Conference on Control Applications, Dayton, OH, 1992, pp. 962–963 (1992)
2. Chai, T., Joe Qin, S., Wang, H.: Optimal operational control for complex industrial processes. Ann. Rev. Control **38**(1), 81–92 (2014)
3. Li, Y., Ang, K.H., Chong, G.C.: Patents, software, and hardware for PID control: an overview and analysis of the current art. IEEE Control Syst. Mag. **26**(1) 42–54 (2006)
4. Ziegler, J.G., Nichols, N.B.: Optimum settings for automatic controllers. Trans. ASME **64**(11), 759–768 (1942)
5. Meshram, P.M., Kanojiya, R.G.: Tuning of PID controller using Ziegler-Nichols method for speed control of DC motor. In: IEEE-International Conference on Advances in Engineering, Science and Management, pp. 117–122 (2012)
6. Aström, K.J, Hägglund, T.: PID Controllers: Theory, Design, and Tuning, pp. 1–354. Instrument Society of America, Research Triangle Park (2012)
7. Zhang, P., Yuan, M., Wang,H.: Self-tuning PID based on adaptive genetic algorithms with the application of activated sludge aeration process. In: 6th World Congress on Intelligent Control and Automation, Dalian, pp. 9327–9330 (2006)
8. Aström, K.J., Hägglund, T., Hang, C.C., Ho, W.O.: Automatic tuning and adaptation for PID controllers - a survey. Control Eng. Pract. **1**(4), 699–714 (1993)
9. Kim, K., Schaefer, R.C.: Tuning a PID controller for a digital excitation control system. IEEE Trans. Ind. Appl. **41**(2), 485–492 (2005)
10. Ho, W.K., Xu, W.: PID tuning for unstable processes based on gain and phase-margin specifications. IEE Proc. Control Theory Appl. **145**(5), 392–396 (1998)
11. Paraskevopoulos, P.N., Pasgianos, G.D., Arvanitis, K.G.: PID-type controller tuning for unstable first order plus dead time processes based on gain and phase margin specifications. IEEE Trans. Control Syst. Technol. **14**(5). 926–936 (2006)
12. Jinggong, L., Yali, X., Donghai, L.: Calculation of PI controller stable region based on D-partition method. In: International Conference on Control, Automation and Systems (ICCAS) 2010, Gyeonggi-do, pp. 2185–2189 (2010)
13. Hwang, C., Hwang, J.-H., Leu, J.-F.: Tuning PID controllers for time-delay processes with maximizing the degree of stability. In: 5th Asian Control Conference (IEEE Cat. No.04EX904), pp. 466–471 (2004)
14. Chen, D., Seborg, D.E.: Design of decentralized PI control systems based on Nyquist stability analysis. J. Proc. Control **13**(1), 27–39, (2003)
15. Vrána, S., Šulc, B.: PID contoller autotuning based on nonlinear tuning rules. In: 12th International Carpathian Control Conference (ICCC), pp. 443–446 (2011)
16. Cohen G.: Theoretical consideration of retarded control. Trans. Asme. **75**, 827–34 (1953)
17. Ho, W.K., Gan, O.P., Tay, E.B., Ang, E.L.: Performance and gain and phase margins of well-known PID tuning formulas. IEEE Trans. Control Syst. Technol. **4**(4), 473–477 (1996)
18. Rivera, D.E., Morari, M., Skogestad, S.: Internal model control: PID controller design. Ind. Eng. Chem. Proc. Des. Dev. **25**(1), 252–265 (1986)
19. Tan, W.: Unified tuning of PID load frequency controller for power systems via IMC. IEEE Trans. Power Syst. **25**(1), 341–350 (2010)
20. Grassi, E., et al.: Integrated system identification and PID controller tuning by frequency loop-shaping. IEEE Trans. Control Syst. Technol. **9**(2), 285–294 (2001)
21. Ogata, K., Yang, Y.: Modern Control Engineering, vol. 17. Pearson, Upper Saddle River (2010)
22. Padula, F., Visioli, A.: Optimal tuning rules for proportional-integral-derivative and fractional-order proportional-integral-derivative controllers for integral and unstable processes. IET Control Theory Appl. **6**(6), 776–8 (2012)
23. Reynoso-Meza, G., Sanchis, J., Blasco, X., Martínez, M.: Algoritmos evolutivos y su empleo en el ajuste de controladores del tipo PID: Estado Actual y Perspectivas. RIAI Rev. Iberoam. Autom. Inform. Ind. **10**(3) 251–68 (2013)

24. Killingsworth, N.J., Krstic, M.: PID tuning using extremum seeking: online, model-free performance optimization. IEEE Control Syst. Mag. **26**(1), 70–79 (2006)
25. Blondin, M.J., Sicard, P.: ACO based controller and anti-windup tuning for motion systems with flexible transmission. In: 2013 26th IEEE Canadian Conference on Electrical and Computer Engineering (CCECE), pp. 1–4. IEEE, Piscataway (2013)
26. Kircher, Jr., et al.: System, method and article for controlling the dispense of insulin, US 9,056,168 B2 (2015)
27. Blondin, M.J., Sanchis Sáez, J., Panos, M.P.: Control Engineering from Classical to Intelligent Control Theory—An Overview, Computational Intelligence and Optimization Methods for Control Engineering. Springer, Cham (2019)
28. Smith, B.: Classical vs. Intelligent Control, EN9940 Special Topics in Robotics, 9645953 (2002)
29. He, J.B., Wang, Q.G., Lee, T.H.: PI/PID controller tuning via LQR approach. Chem. Eng. Sci. **55**(13), 2429–2439 (2000)
30. Kafi, M.R., Chaoui, H., Miah, S., Debilou, A.: Local model networks based mixed-sensitivity H-infinity control of CE-150 helicopters. Control Theory Technol. **15**(1), 34–44 (2017)
31. Blondin, M.J., Pardalos, P.M., Sáez, J.S.: Computational Intelligence and Optimization Methods for Control Engineering, Springer Optimization and Its Applications, pp. 1–355. Springer, Berlin (2019)
32. Du, F., Dong, Q.-Y., Li, H.-S.: Truss structure optimization with subset simulation and augmented Lagrangian multiplier method. Algorithms **10**(4), 128 (2017)
33. Xiong, N., Molina, D., Ortiz, M.L., Herrera, F.: A walk into metaheuristics for engineering optimization: principles, methods and recent trends. Int. J. Comput. Intell. Syst. **8**(4), 606–636 (2015)
34. Hussain, K., Salleh, M.N.M., Cheng, S., Shi, Y.: Metaheuristic research: a comprehensive survey. Artif. Intell. Revi. **52**(4), 2191–2233 (2019)

Chapter 2
Controller Tuning by Metaheuristics Optimization

2.1 Historical Evolution of Metaheuristics

This section presents an overview of the progression of metaheuristics from its beginning to date. The *historic milestone* of the development of metaheuristics is presented, allowing the reader capture the essence of metaheuristics evolution and its history. Since the earliest centuries, practical methods, referred to as heuristics, have solved optimization problems. These methods do not guarantee to reach optimal solutions, but their main advantage over exact methods is their low calculation costs. Calculation cost remains a crucial aspect in optimization methods, even though technological advances have enabled the development and the use of sophisticated computational tools. Metaheuristics have a higher procedure level than heuristics, i.e., they are suitable for a wide spectrum of optimization problems without requiring significant changes. Indeed, metaheuristics make few assumptions about the problem to solve, which makes their algorithmic framework problem-independent.

The term *metaheuristic* has been created recently, and its definition has not reached agreement among the scientific community [265]. One definition can be as follows: *a metaheuristic is a high-level abstraction of established sets of mathematical rules with some randomness that leads a search process to find (near)-optimal solutions to a problem.* As a result, it is almost impossible to identify the first metaheuristics. Nonetheless, it is known that the 1960s and 1970s paved the way to metaheuristics; evolutionary strategies and evolutionary programming techniques were developed as search techniques to solve optimization problems [92]. Along the same lines, inspired by Darwin's evolution theory and natural selection, a search method, called genetics algorithm (GA), has been proposed [61]. This method performs mathematical operations such as crossover and mutations on potential solutions to solve optimization problems. Since its publication, GA has been applied to thousands of systems of various applications, from economics to

© Springer Nature Switzerland AG 2021
M. J. Blondin, *Controller Tuning Optimization Methods for Multi-Constraints and Nonlinear Systems*, SpringerBriefs in Optimization,
https://doi.org/10.1007/978-3-030-64541-0_2

engineering. Many modified versions of GA have been proposed. Each of them possesses differences with the intent of improving the search process to reach better solutions. These techniques have lain the foundations of generic optimization methods that could optimize almost any problem "effectively."

The following decades, the 1980s and 1990s, have been innovative. Many optimization algorithms have been developed, most of which use metaphors. Among the first metaheuristics after GA, simulated annealing (SA), an algorithm based on the heating and cooling processes of material, was proposed [156]. Another approach similar to GA [61] is the artificial immune system (AIS). This algorithm mimics the immune system's behavior, in which the latter is capable of learning, memorizing, and pattern recognizing [83]. Besides, animal and insect behavior studies have been a source of inspiration to develop many more optimization methods. For instance, ant colonies tend to choose the shortest path between their nest and a food source. This ant behavior has been metaphorically employed to conceive an optimization algorithm called ant colony optimization (ACO) [68]. ACO was first applied to find the optimal path in graphs. No longer after ACO, the social behavior of bird flocking or fish schooling has inspired the development of the particle swarm optimization (PSO) algorithm [153]. Since their first publication, several variants of ACO and PSO have been proposed to achieve better results. The 1980s and the 1990s have built the groundwork of devising optimization algorithms based on observations/behaviors of nature that are frameworks to create metaheuristics. From the 2000s, metaheuristics gain in popularity, and researchers have been developing a tremendous number of optimization methods. This kind of optimization technique has grown tremendously among the scientific community, creating the metaheuristics field itself. Conferences on the topic have been created, such as the Genetic and Evolutionary Computation Conference, the International Conference on Optimization, Metaheuristics and Machine Learning, the International Conference on Metaheuristics, as well as the IEEE World Congress on Computational Intelligence. As new algorithms are published every month, it is nearly impossible to survey them all. However, the next subsection intends to report them as many as possible.

2.2 Metaheuristics Survey

Table 2.1 presents an exhaustive survey of metaheuristics and their year of publication and their first application.

For most metaheuristics, their first application that validated their efficiency is benchmark functions. CEC benchmark functions refer to the benchmark functions proposed in the Congress on Evolutionary Computation competition, e.g., CEC05 benchmark functions means that the algorithm was tested on some of the benchmark functions proposed in the 2005 Congress on Evolutionary Computation. Only ten metaheuristics were specifically proposed for controller tuning. Although tough metaheuristics are problem-independent, research on developing metaheuristics for

Table 2.1 Meta-survey of metaheuristics

Optimization algorithm	Ref.	Year	First application
Across neighborhood search	[301]	2016	Benchmark functions
Adaptive dimensional search	[113]	2015	Discrete truss sizing problems
Adaptive gravitational search algorithm	[229]	2012	Fuzzy controlled servo systems
Adaptive evolution	[288]	2009	Benchmark functions
Adaptive social behavior optimization	[263]	2013	Benchmark problems
Adaptive spiral dynamics algorithm	[216]	2013	Benchmark functions
			PD controller tuning of a flexible manipulator system
African buffalo optimization	[220]	2016	Global optimization
			Traveling salesman problem
African wild dog algorithm	[268]	2012	Engineering problems
Ageist spider monkey optimization algorithm	[256]	2016	Benchmark functions
Alienated ant algorithm	[20]	2010	Grid jobs scheduling
Amoeboid organism algorithm	[322]	2013	0–1 knapsack problems
Anarchic society optimization	[259]	2012	PID controller tuning for AVR system
Animal migration optimization algorithm	[166]	2014	Benchmark functions
Ant colony optimization	[67]	2004	Combinatorial optimization problems
Ant colony system	[66]	1997	Traveling salesman problem
Ant lion optimizer	[186]	2015	Benchmark functions
AntStar	[80]	2015	Shortest-path problems
Artificial algae algorithm	[286]	2015	CEC05 benchmark functions
			Pressure vessel design
Artificial bee colony	[135]	2007	Constrained optimization problems
Artificial chemical reaction optimization algorithm	[8]	2011	Multiple-sequence alignment
			Data mining
			Benchmark functions
Artificial cooperative search algorithm	[49]	2013	Benchmark functions
			CEC05 and CEC08 benchmark functions
			Welded beam design
			Pressure vessel design
			Tension/compression spring design problem
			Speed reducer design
Artificial ecosystem algorithm	[5]	2014	Traveling salesman problem
Optimization algorithm	Ref.	Year	First application
Artificial feeding birds	[162]	2019	Benchmark functions
			Traveling salesman problem
			Optimization of rainbow boxes
Artificial fish swarm optimization algorithm	[165]	2003	Optimization problems

(continued)

Table 2.1 (continued)

Optimization algorithm	Ref.	Year	First application
Artificial root foraging algorithm	[176]	2015	CEC13 and CEC14 benchmark functions
Asexual reproduction optimization	[82]	2010	Benchmark functions
Atmosphere clouds optimization	[97]	2013	Benchmark functions
Bacterial colony optimization	[219]	2012	Benchmark functions
Bacterial foraging optimization algorithm	[225]	2002	Distributed optimization and control
Backtracking search optimization algorithm	[50]	2013	Boundary-constrained benchmark problems
			Constrained real world benchmark problems
Bat-inspired algorithm	[309]	2010	Benchmark functions
Bayesian optimization algorithm	[227]	1999	Benchmark functions
Bean optimization algorithm	[323]	2013	Benchmark functions
Beehive	[211]	2009	Benchmark functions
Bees algorithm	[319]	2013	Benchmark functions
Binary artificial algae algorithm	[25]	2018	Wind turbine placement problem
Big bang–big crunch	[77]	2006	Benchmark functions
Binary bat algorithm	[193]	2014	Benchmark functions
Binary cat swarm optimization	[253]	2015	Benchmark optimization problems and 0–1 knapsack problem
Binary magnetic optimization algorithm	[190]	2012	Benchmark functions
Binary spider monkey optimization algorithm	[264]	2016	Thinning of concentric
			Circular antenna arrays
Binary whale optimization algorithm	[240]	2019	Profit-based unit commitment
			Problems in competitive electricity markets
Bird mating optimizer	[16]	2014	Benchmark functions
Bird swarm algorithm	[183]	2016	Benchmark problems
Biogeography-based optimization	[262]	2018	Benchmark functions
Biology migration algorithm	[325]	2019	Benchmark functions
			Real engineering problems
Black hole	[115]	2013	Data clustering
Blind, naked mole-rats optimization	[274]	2013	Data clustering
Bottlenose dolphin optimization	[157]	2015	Traveling salesman problem
Brain storm optimization algorithm	[261]	2011	Benchmark functions
Bull optimization algorithm	[90]	2015	Benchmark functions
Bumble bees mating optimization	[179]	2010	Benchmark functions
Butterfly optimization algorithm	[14]	2018	Benchmark functions
Camel algorithm	[122]	2016	Benchmark functions
Car tracking optimization algorithm	[44]	2018	Benchmark functions
Cat swarm optimizer	[47]	2006	Benchmark functions
Center of mass optimization	[102]	2018	Seismic topology optimization
Central force optimization	[173]	2015	Benchmark functions

Table 2.1 (continued)

Optimization algorithm	Ref.	Year	First application
Chaotic krill herd optimization	[246]	2014	Benchmark functions
Charged system search	[147]	2010	Benchmark functions and engineering design problems
Chemical reaction optimization	[161]	2009	Quadratic assignment problem
			Resource-constrained project scheduling problem
			Channel assignment problem
Chicken swarm optimization	[182]	2014	Benchmark functions
			Speed reducer design
Clonal selection algorithm	[57]	2005	Toy problems
			Pattern recognition
			Benchmark functions
			Protein structure prediction–HP model
Cloud particles differential evolution algorithm	[167]	2015	Benchmark functions and CEC13 problems
Coalition-based metaheuristic	[181]	2010	Vehicle routing problem
Colliding bodies optimization	[146]	2014	Design of a pressure vessel
			Design of a tension/compression spring
			Weight minimization of the 120-bar truss dome
			Design of forth truss bridge
Colonial competitive algorithm	[98]	2008	PID controller for a typical distillation column process
Competition over resources	[203]	2014	Benchmark functions
Comprehensive learning particle swarm optimizer	[170]	2006	Benchmark functions
Consultant-guided search	[124]	2010	Traveling salesman problem
Contour gradient optimization	[304]	2013	Benchmark functions
Cooperative group optimization with ants	[307]	2017	Traveling salesman problem
Cooperative particle swarm optimization	[287]	2004	Benchmark functions
Coral reefs optimization algorithm	[244]	2014	Benchmark functions
			Traveling salesman problem
			Mobile network deployment problem
			Design of off-shore wind farms
Coyote optimization algorithm	[228]	2018	CEC05 and CEC15 Benchmark functions
Creativity-oriented optimization algorithm	[85]	2015	CEC-2013 real-parameter optimization benchmark functions
Cricket algorithm	[40]	2016	Three-bar truss
			Himmelblau's function
			Spring design problem
			Speed reducer

(continued)

Table 2.1 (continued)

Optimization algorithm	Ref.	Year	First application
Cross-entropy method	[60]	2005	Max-cut problem
			Traveling salesman problem
Crow search algorithm	[17]	2016	Three-bar truss design problem
			Pressure vessel design problem
			Tension/compression spring design problem
			Welded beam design problem
			Gear train design problem
			Belleville spring design problem
			Benchmark functions
Crystal energy optimizer	[86]	2016	Traveling salesman problem
Cuckoo optimization algorithm	[235]	2011	Benchmark functions
			PID controller designed for a MIMO chemical system
Cultural algorithm approach	[131]	1999	Nonlinear constraint optimization problem [91]
Cuttlefish algorithm	[72]	2013	Benchmark functions
Curved space optimization	[200]	2012	Benchmark functions
Cyber swarm algorithm	[317]	2010	Benchmark functions
Cyclical parthenogenesis algorithm	[149]	2017	Benchmark functions
			Truss design problems
Dialectic search	[133]	2009	Costas arrays problems
			Magic squares problems
			Benchmark functions
Differential evolution	[266]	1997	Benchmark functions
Differential search algorithm	[48]	2012	Transforming the geocentric Cartesian coordinates
			Benchmark functions
Discrete firefly algorithm	[26]	2018	Constraint satisfaction problems
Dolphin echolocation	[143]	2013	Truss structures
			Structural optimization
			Frame structures
Dragonfly algorithm	[189]	2015	Benchmark functions
			Submarine's propeller
Duelist algorithm	[27]	2016	Benchmark functions
Eagle strategy	[312]	2010	Benchmark functions
Earthworm optimization algorithm	[294]	2018	Benchmark functions
			Test-sheet composition
Ecogeography-based optimization	[329]	2014	Benchmark functions
			Emergency airlift problem
Egyptian vulture optimization	[271]	2013	0–1 Knapsack problems
Election algorithm	[76]	2015	Mathematical benchmark examples
Electromagnetic field optimization	[4]	2016	CEC14 benchmark functions

Table 2.1 (continued)

Optimization algorithm	Ref.	Year	First application
Electromagnetism metaheuristic	[89]	2013	Betweenness problem
Electro-search algorithm	[272]	2017	Benchmark functions
			Nonlinear constrained optimization problem [121]
			Industrial chemical plant
Elephant herding optimization	[293]	2016	Benchmark functions
			Optimal control of a nonlinear stirred tank reactor
			Dynamic economic dispatch problem
Elephant search algorithm	[64]	2015	Benchmark functions
Elitist stepped distribution algorithm	[10]	2017	Benchmark functions
			Welded beam design
			Pressure vessel design
			Stepped cantilever beam design
Emperor penguins colony	[112]	2019	Benchmark functions
Enhanced best performance algorithm	[46]	2015	Just-in-time scheduling problem
Equilibrium optimizer	[81]	2019	Benchmark functions
			Pressure vessel design
			Welded beam design
			Tension/compression spring design
Evolutionary multi-objective immune algorithm	[278]	2008	Benchmark problems
Exchange market algorithm	[103]	2014	Benchmark functions
Extremal optimization	[35]	2000	Graph partitioning
			Traveling salesman problem
Farmland fertility	[258]	2018	Benchmark functions
FIFA World Cup competitions based	[238]	2005	Benchmark functions
			PID controller tuning for the AVR system
Find-fix-finish-exploit-analyze	[140]	2019	Pressure vessel design
			Simply supported 37-bar planar truss structure
			Speed reducer design
			Centrifugal pump design
			Gear train design
			Parameter estimation of a frequency-modulated synthesizer
			Benchmark functions
Firefly algorithm	[310]	2013	Design of a tension/compression spring
Fireworks algorithm	[277]	2010	Benchmark functions
Fish electrolocation optimization	[111]	2017	Benchmark functions
			Cost-based reliability enhancement in radial distribution system

(continued)

Table 2.1 (continued)

Optimization algorithm	Ref.	Year	First application
Fisherman search procedure	[9]	2014	Benchmark functions
Fish swarm algorithm	[164]	2002	Benchmark functions
Flow regime algorithm	[273]	2019	Benchmark functions
			Heat wheel
			Horizontal axis marine current turbine
Flower pollination algorithm	[311]	2012	Benchmark functions
			Design optimization
Flying elephants algorithm	[305]	2016	Clustering problems
Foraging agent swarm optimization	[21]	2014	Clustering-data set
			The Fermat–Weber problem
			Hub location problems
Forest optimization algorithm	[100]	2014	Benchmark functions
			Feature weighting
Fruit fly algorithm	[224]	2012	General regression neural network
			General regression neural network
			Multiple regression
Gaining sharing knowledge	[202]	2019	CEC2017 Benchmark functions
			Real world optimization problems of IEEE-CEC2011 competition
Galaxy-based search algorithm	[252]	2011	Principal components analysis estimation
Galactic swarm optimization algorithm	[213]	2016	Benchmark functions
Gases Brownian motion optimization	[2]	2013	Benchmark functions
			Lennard–Jones potential problem
			Tersoff potential function minimization problem
Genetic algorithm	[61]	1975	Benchmark functions
Generalized acquisition of recurrent links	[12]	1994	Williams' trigger problem
			Inducing regular languages
			The ant problem
Global simplex optimization	[137]	2012	Benchmark functions
Glowworm swarm optimization	[158]	2009	Benchmark examples
Golden ball	[222]	2014	Traveling salesman problem
			Capacitated vehicle routing problem
Gradient evolution algorithm	[159]	2015	Benchmark functions
Gradient gravitational search	[59]	2015	Benchmark functions
			Potential energy of 2D and 3D off-lattice protein models
Grasshopper optimization algorithm	[247]	2017	CEC05 Benchmark functions
			Benchmark functions
			Three-bar truss design problems
			Cantilever beam design problem
			52-bar truss design

Table 2.1 (Continued)

Optimization algorithm	Ref.	Year	First application
Gravitational search algorithm	[237]	2009	Benchmark functions
Great salmon run	[207]	2012	Benchmark optimization problems
Greedy randomized adaptive search procedures	[87]	1989	Integer programming problems [93]
Green herons optimization algorithm	[270]	2013	Traveling salesman problem
Grenade explosion method	[6]	2010	Traveling salesman problem
Grey wolf optimizer	[194]	2014	CEC05 benchmark functions
			Benchmark functions
			Welded beam design
			Pressure vessel design
			Optical buffer design
			Tension/compression spring design
Group search optimizer	[117]	2009	Benchmark functions
			Artificial neural networks training
Harmony search	[99]	2001	Traveling salesman problem
			Least-cost pipe network design problem
Harris hawks optimization	[118]	2019	CEC05 Benchmark functions
			Benchmark functions
			Three-bar truss design problems
			Tension/compression spring problem
			Pressure vessel design problem
			Welded beam design
			Multi-plate disc clutch brake
			Rolling element bearing design problem
Heuristic Kalman algorithm	[284]	2009	Benchmark functions
			Welded beam design
			Robust PID controller tuning
Heart optimization algorithm	[116]	2014	Data clustering
Heat transfer search	[249]	2017	Half car ride model design
Honey bee algorithm	[215]	2004	Server allocation
Honey bees mating optimization	[109]	2006	Benchmark functions
			Single reservoir operation optimization
Hoopoe heuristic	[74]	2012	Benchmark functions
Human-inspired algorithm	[321]	2009	Benchmark functions
Hunting search	[221]	2010	Benchmark functions
Hurricane-based optimization algorithm	[239]	2014	Benchmark functions
Hydrological cycle algorithm	[297]	2017	Benchmark functions
Hyper-spherical search	[136]	2014	Benchmark functions
Imperialist competitive algorithm (ICA)	[18]	2007	Benchmark functions
Improved bees algorithm	[255]	2015	Dynamic economic dispatch

(Continued)

Table 2.1 (Continued)

Optimization algorithm	Ref.	Year	First application
Improved magnetic charged system search	[151]	2015	10-Bar planar truss structure
			A 52-bar planar truss
			72-bar spatial truss
			A 120-bar dome shaped truss problem
Improved quantum-behaved particle swarm optimization	[306]	2008	Benchmark functions
Improved whale optimization algorithm	[38]	2019	Benchmark functions
Intelligent water drops algorithm	[251]	2008	Multiple knapsack problem
Interactive search algorithm	[206]	2018	Benchmark functions
			Gear system design
			Pressure vessel design
			Welded beam design
			Tension/compression spring design
			Speed reducer design problem
Interior search algorithm	[94]	2014	Benchmark functions and engineering problems
Invasive tumor growth optimization	[279]	2015	CEC05, CEC08, CEC10 benchmark functions
			Support vector machine parameter optimization problem
Invasive weed optimization	[180]	2006	Benchmark functions
			Controller optimization
Ions motion algorithm	[130]	2015	Benchmark functions
JADE algorithm	[320]	2009	Benchmark functions
Jaguar algorithm	[43]	2015	Benchmark functions
Japanese tree frogs	[120]	2012	Distributed graph coloring problems
Jenga-inspired optimization algorithm	[163]	2012	Energy-efficient coverage problem in wireless sensor networks
Joint operations algorithm	[269]	2016	CEC12 benchmark functions
			Solving systems of linear equations
			Parameter optimization for polynomial fitting problem
			CEC11 real-life optimization problems
Kaizen programming	[62]	2014	Benchmark functions
			Nguyen benchmark functions
Key cutting algorithm	[232]	2009	9-number puzzle problem
Killer whale algorithm	[28]	2017	Benchmark functions
Kinetic gas molecule optimization	[199]	2014	Benchmark functions
Krill herd algorithm	[95]	2012	Benchmark functions
Leaders and followers	[108]	2015	
League championship algorithm	[138]	2009	Benchmark functions

Table 2.1 (Continued)

Optimization algorithm	Ref.	Year	First application
Levy flight artificial bee colony algorithm	[257]	2016	Benchmark functions
			Pressure vessel design
			Lennard–Jones
			Frequency-modulated sound wave
			Compression spring
			Welded beam design
Life choice-based optimizer	[154]	2019	Benchmark functions
			Pressure vessel design
			Cantilever beam design
Lifecycle-based swarm optimization	[260]	2013	Benchmark functions
Lightning attachment procedure optimization	[218]	2017	
Lion optimization algorithm	[316]	2016	Benchmark functions
Lion pride optimizer	[291]	2012	Benchmark functions
Locust swarm algorithm	[56]	2015	Benchmark functions
Macroevolutionary algorithms	[178]	1999	Benchmark functions
Magnetic charged system search	[150]	2013	Benchmark functions
			Tension/compression spring design problem
			Welded beam design
			Pressure vessel design
Magnetotactic bacteria optimization	[198]	2013	Benchmark functions
Magnetic optimization algorithm	[192]	2011	3-bit XOR and function approximation benchmarks problems
Marriage in honey bees optimization algorithm	[1]	2001	Propositional satisfiability problems
Mean–variance mapping optimization	[214]	2011	Optimal reactive power dispatch problems
Melody search algorithm	[15]	2013	Benchmark functions
Migrating birds optimization algorithm	[71]	2012	Quadratic assignment problems
Mine blast algorithm	[243]	2013	Benchmark functions
			Three-bar truss design problem
			Pressure vessel design
			Tension/compression spring design
			Welded beam design
			Speed reducer design
			Gear train design
			Belleville spring design
			Rolling element bearing design
Modified distributed bees algorithm	[283]	2013	Multi-sensor task allocation

(Continued)

Table 2.1 (Continued)

Optimization algorithm	Ref.	Year	First application
Modified backtracking search algorithm	[295]	2018	Benchmark functions
			Three-bar truss design problem
			Pressure vessel design
			Tension/compression spring design
			Welded beam design
			Speed reducer design
Monarch butterfly optimization	[296]	2019	Benchmark functions
Monkey algorithm	[327]	2008	Benchmark functions
Monkey search algorithm	[209]	2007	Lennard–Jones and Morse clusters
			The tube model
Mosquito host-seeking algorithm	[84]	2013	Traveling salesman problem
Moth-flame Optimization	[187]	2015	CEC05 Benchmark functions
			Benchmark functions
			Welded beam design problem
			Gear train design problem
			Three-bar truss design problem
			Cantilever beam design problem
			Pressure vessel design problem
			I-beam design problem
			Tension/compression spring design
			15-bar truss design
			Marine propeller design
Moth search algorithm	[290]	2018	Benchmark functions
			CEC05 benchmark functions
			CEC11 real world problems
Mouth brooding fish algorithm	[126]	2017	
MOX algorithm	[210]	2011	Benchmark functions
Multi-objective PSO	[52]	2002	Benchmark functions
Multi-objective root system growth optimizer	[177]	2017	Multi-objective benchmarks
Multi-verse	[195]	2016	Benchmark functions
			CEC05 benchmark functions
			Welded beam design
			Gear train design
			Three-bar truss design
			Pressure vessel design
			Cantilever beam design
Musical composition inspired algorithm	[205]	2014	Benchmark functions
Mussels wandering optimization	[11]	2013	Benchmark functions
Natural aggregation algorithm	[175]	2016	Benchmark functions
Non-dominated sorting genetic algorithm II	[63]	2000	Benchmark functions

Table 2.1 (Continued)

Optimization algorithm	Ref.	Year	First application
Nuclear reaction optimization	[300]	2019	Benchmark functions
			CEC18 test suite
			Welded beam design problem
			Pressure vessel design
			Tension/compression spring design
Open source development model algorithm	[110]	2016	Benchmark functions
			CEC14 benchmark functions
Optics inspired optimization	[139]	2015	Benchmark functions
			CEC05 benchmark functions
			Bi-objective optimization of a centrifuge pump
Optimization booster algorithm	[223]	2019	Benchmark functions
			Pressure vessel design
			Bi-objective optimization of a centrifuge pump
			Welded beam design
			Traveling salesman problem
Owl search algorithm	[127]	2018	Benchmark functions
Paddy field algorithm	[230]	2009	Benchmark functions
Parliamentary optimization algorithm	[36]	2009	Benchmark functions
Particle migration	[96]	2012	Benchmark functions
Particle swarm optimization	[153]	1995	Benchmark functions
			Train a neural network to classify the Fisher Iris data set
Passing vehicle search	[248]	2016	Pressure vessel design
			Spring design optimization
			Welded beam design
			Speed reducer problem
			Bearing design optimization
			Multi-plate disc clutch brake optimization
			Step-cone pulley
			Belleville spring
			Robot gripper
			Hydrostatic thrust bearing
			Planetary gear train design optimization
			Stiffened welded shell design optimization
			Four-stage gear box

(Continued)

Table 2.1 (Continued)

Optimization algorithm	Ref.	Year	First application
Pathfinder algorithm	[315]	2019	Benchmark functions
			Pressure vessel design
			Tension/compression spring design
			Welded beam design
			Cantilever beam design
			Optimal placement and sizing of renewable energy sources
Particle fish swarm algorithm	[303]	2012	Benchmark functions
Penguins search optimization algorithm	[101]	2013	Benchmark functions
Photosynthetic learning algorithm	[212]	1998	Traveling salesman problem
Pigeon optimization	[69]	2014	Air robot path planning
Pity beetle algorithm	[134]	2018	Benchmark functions
			CEC14 benchmark functions
POPMUSIC	[275]	2002	Large centroid clustering
			Balancing mechanical parts
Prey–predator algorithm	[282]	2015	Benchmark functions
Radial movement optimization	[234]	2014	Benchmark functions
Raindrop algorithm	[299]	2013	Optimization problem
Rainfall optimization algorithm	[132]	2017	Benchmark functions
			Economic dispatch optimization problem
Raven roosting optimization algorithm	[39]	2016	Benchmark functions
Ray optimization	[145]	2012	Benchmark functions
			Tension/compression spring design problem
			Welded beam design problem
			25-bar spatial truss
Real-coded GA	[152]	1997	Controller tuning
Reinforced quantum-behaved PSO	[114]	2016	LQR tuning for inverted pendulum and flight landing system
River formation dynamics	[233]	2007	Traveling salesman problem
Root growth algorithm	[324]	2014	Benchmark functions
Runner-root algorithm	[185]	2015	Benchmark functions
			CEC05 benchmark functions
Sailfish optimizer	[250]	2019	Benchmark functions
			I-beam design problem
			Welded beam design
			Gear train design
			Three-bar truss design
			Circular antenna array design
Salp swarm algorithm	[196]	2017	Benchmark functions
			CEC09 benchmark functions
			Three-bar truss design

Table 2.1 (Continued)

Optimization algorithm	Ref.	Year	First application
			Welded beam design
			I-beam design problem
			Cantilever beam design problem
			Tension/compression spring design
			Two-dimensional airfoil design
			Marine propeller design
Scatter search	[105]	2003	Nonlinear optimization problems
Search group algorithm	[107]	2015	10-bar plane truss size optimization
			37-bar plane truss size and shape optimization
			11- and 15-bar plane truss size, shape, and topology optimization
			20- and 24-bar plane truss size and topology optimization
Seed disperser ant algorithm	[42]	2015	Benchmark functions
Seed throwing optimization	[298]	2009	Benchmark functions
Seeker optimization algorithm	[58]	2006	Benchmark function
			Reactive power dispatch—IEEE 57-bus and 118-bus Power System
Self-defense mechanisms of plants	[41]	2018	Benchmark functions
			CEC15 benchmarks functions
Seven-spot ladybird optimization	[292]	2013	Benchmark functions
Shark smell optimization	[3]	2016	Benchmark functions
			CEC13 benchmarks functions
			PID controller tuning for load frequency control
Shuffled artificial bee colony algorithm	[254]	2017	CEC06 benchmark functions
			Reactor network design
			Network design of a heat exchanger
			Separation network synthesis problem
			Optimal operation of alkylation unit
			Optimal distribution policy to minimize a company total cost
Shuffled complex evolution	[70]	1993	Benchmark functions
Shuffled frog leaping algorithm	[79]	2006	Traveling salesman problem
			Calibrate a steady-state hypothetical groundwater model
			Simpleton25
			Simpleton50
			Cutting stock
			Trim loss
			Gear problem

(Continued)

Table 2.1 (Continued)

Optimization algorithm	Ref.	Year	First application
Simple adaptive climbing	[289]	2014	Benchmark functions
Simplifying particle swarm optimization	[226]	2010	Artificial neural networks weight optimization problems
Simulated annealing	[156]	1983	Graph partitioning
			Traveling salesman problem
Sine–cosine algorithm	[188]	2016	Benchmark functions
			Airfoil design
Social emotional optimization algorithm	[308]	2010	A typical nonlinear programming problem
Soccer league competition algorithm	[204]	2014	Water distribution networks
Social network-based swarm optimization	[171]	2015	Benchmark functions
Social spider algorithm	[129]	2015	CEC13 and CEC14 benchmark functions
Sonar inspired optimization	[285]	2017	Benchmark functions
Spider monkey optimization	[7]	2016	Linear antenna array synthesis
			E-shaped patch antenna design
Spiral optimization algorithm	[276]	2011	Benchmark function
Spiritual search algorithm	[231]	2018	Model identification of DC motor
			PID and PIDA controller tuning
Squeaky wheel optimization	[51]	1999	Fiber-optic production line scheduling problem
			Graph coloring problem
Squirrel search algorithm	[128]	2019	Benchmark functions
			CEC14 benchmark functions
			2DOFPI controller for temperature control
States of matter search optimization algorithm	[55]	2014	Benchmark functions
			GECCO benchmark functions
Strawberry plant optimization algorithm	[184]	2014	Benchmark functions
			Controller tuning for DC motor
Stochastic diffusion search	[217]	1999	Locating best instantiation of the object in the search space
Stochastic focusing search	[318]	2009	Benchmark functions
Stochastic fractal search	[245]	2015	Benchmark functions
			CEC10 benchmark functions
			Tension/compression spring design
			Pressure vessel design
			Welded beam design
Symbiotic organism search	[45]	2014	Benchmark functions
			Cantilever beam design
			Minimize I-beam vertical deflection
			A 15-bar planar truss structure
			52-bar planar truss structure

Table 2.1 (Continued)

Optimization algorithm	Ref.	Year	First application
Swarm flow algorithm	[241]	2014	Smart energy dispatching
Swarm-inspired projection based on doves	[267]	2009	Estimating the number of clusters existing in the data set
Tabu search algorithm	[104]	1989	Zero-one knapsack problem
			Scheduling problem for distributing workloads among machines
			Load-balancing channel assignment
			Clustering problem used in space planning and architectural design
			Large scale employee scheduling problem
Teaching–learning-based optimization	[236]	2011	Benchmark functions
			Gear train
			Tension/compression spring design
			Pressure vessel design
			Welded beam design
			Multiple disc clutch brake
			Robot gripper
			Step-cone pulley
			Hydrodynamic thrust bearing
			Rolling element bearing
			Belleville spring
Team arrangement heuristic algorithm	[19]	2019	Benchmark functions
			CEC05 benchmark functions
			Heat wheel optimization problem
			Horizontal axis tidal current turbine problem
Thermal exchange optimization	[142]	2017	Benchmark functions
			Welded beam design
			Tension/compression spring design
			Stepped cantilever beam design
			Pressure vessel design
Tree seed algorithm	[155]	2015	Benchmark functions
			Image thresholding
Tug of war	[148]	2016	Benchmark functions
			Welded beam design
			Tension/compression spring design
			Planar 10-bar truss structure design
			Spatial 25-bar truss structure design
			Spatial 72-bar truss structure design
			120-bar dome truss
Two lbests multi-objective particle swarm optimization	[326]	2011	CEC07 benchmark problems

(continued)

Table 2.1 (Continued)

Optimization algorithm	Ref.	Year	First application
Vapor–liquid equilibrium based algorithm	[54]	2018	Benchmark functions
			CEC17 benchmark functions
Variable depth search algorithm	[37]	2015	Constraint satisfaction problems
Variable neighborhood search	[197]	1997	Traveling salesman problem
Vibrating particles system	[144]	2017	Sizing optimization of skeletal structures
Virus colony search	[168]	2016	Benchmark functions
			CEC14 benchmark functions
			Pressure vessel design
			Tension/compression spring design
			Welded beam design
Virus optimization algorithm	[169]	2016	Benchmark functions
Volleyball premier league algorithm	[201]	2018	Benchmark functions
			Pressure vessel design
			Tension/compression spring design
			Welded beam design
Vortex search algorithm	[65]	2015	Benchmark functions
War optimization	[53]	2018	69-bus radial distribution system
			476-bus distribution system
Warping search	[106]	2008	Protein structure prediction
Water cycle algorithm	[78]	2012	Benchmark functions
			Speed reducer design
			Three-bar truss design
			Pressure vessel design
			Tension/compression spring design
			Welded beam design
			Rolling element bearing design
			Multiple disk clutch brake design
Water evaporation optimization	[141]	2016	Truss weight minimization
			Planar 10-bar truss design
			Spatial 22-bar truss design
			Spatial 25-bar tower design
			Spatial 72-bar truss design
			120-bar dome truss design
			Planar 200-bar truss
Water flow-like algorithm	[313]	2007	Bin packing problems
Water wave optimization	[328]	2015	CEC14 benchmark functions
			High-speed train scheduling
Weighted artificial fish algorithm	[208]	2015	LQR tuning for inverted cart pendulum
Weighted superposition attraction	[22]	2017	Benchmark functions
			CEC13 benchmark functions

Table 2.1 (Continued)

Optimization algorithm	Ref.	Year	First application
Whale optimization algorithm	[191]	2016	Benchmark functions
			CEC05 benchmark functions
			Tension/compression spring design
			Welded beam design
			Pressure vessel design
			15-bar truss design
			25-bar truss design
			52-bar truss design
Wind driven optimization	[23]	2010	Benchmark functions
			Electromagnetics design problems
Wolf pack algorithm	[302]	2014	Benchmark functions

a specific class of problems, such as control tuning, deserves attention. This survey shows hundreds of metaheuristics in the literature. Most of them have been proposed during the past 10–15 years. In most cases, we can deduce from its name its source of inspiration. In fact, metaheuristics can be classified according to their source of inspiration [191]. The first class is inspired from natural laws of evolution, in which a population evolves by combining the best individuals from one generation to the next such as Genetic algorithms (GA) [61, 191]. The second class relies on physics laws' principles, e.g., simulated annealing (SA) algorithm [156]. Human behavior has motivated the development of algorithms, which represents the third class. For instance, teaching–learning-based optimization (TLBO) [236] and soccer league competition (SLC) algorithm [204] are algorithms inspired by human behavior. Finally, there is a nature-based class. These algorithms imitate group behaviors of individuals, e.g., swarm optimization PSO [153] and ACO [67].

Not only the number of novel metaheuristics skyrocketed during the twenty-first century, but also the number of hybrid algorithms [34]. Hybrid algorithms refer to algorithms that combine two or more algorithms. For instance, it could be a metaheuristic combined to a local search algorithm such as ant colony optimization combined to the Nelder–Mead method (ACO-NM) [32]. It could also be two combined metaheuristics, such as PSO and GA [314]. Chapter 4 refers to those hybrid metaheuristics as it has been a new trend of research. Although the list of metaheuristics is impressive, there are common principles to almost all metaheuristics. The next section presents these principles.

2.3   Metaheuristics Common Principles

Any metaheuristic relies on two major elements during the search process: diversification and intensification. Figure 2.1 illustrates the diversification and intensification processes. The goal of diversification combined with intensification is to convergence to the (near)-optimal solution. The algorithm must avoid converging to a local minimum.

Diversification, also called exploration, is the process of exploring the search space to catch the global optimum's region. This process generates solutions that differ significantly from already visited solutions. *Intensification* or exploitation refers to the use of already collected information to intensify the search in promising areas, i.e., in the vicinity of good solutions. In most cases, diversification and intensification are not performed linearly, but iteratively. Therefore, it is possible that at some point during the search process, the algorithm intensifies its search in local minimum regions. However, through diversification, the algorithm should catch the global optimum's region and, eventually, through the intensification step converge to the global (near)-optimal solution. Ensuring that the found solution is the global (near)-optimum represents one of the biggest optimization challenges.

The key element to reach the (near)-optimum solution is the equilibrium between diversification and intensification. Indeed, insufficient exploration and excessive exploitation could make the algorithm skip the global optimum's region and intensify the search in local minimum regions. Conversely, excessive diversification and insufficient intensification could slow down the algorithm search and make it harder for the algorithm to convergence. In the lexis of optimization, the diversification process means global optimization and intensification and local optimization. This

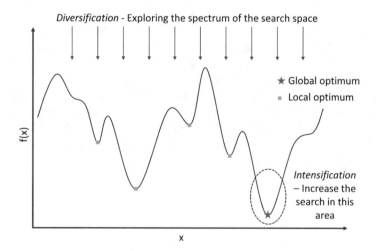

Fig. 2.1 Diversification and intensification processes

combination of global and local optimization within the same algorithm makes the metaheuristics a powerful optimization tool.

A second common component is that any metaheuristic has internal parameters to adjust. The number and function of these parameters depend on the metaheuristics. Assigning values to the different parameters is called parametrization. Parametrization influences metaheuristics performances. Thus, appropriate parametrization is fundamental. For example, in population-based algorithms such as GA, the population's size has to be adequately determined. A too-small population size could make the algorithm converge too soon. At the opposite, a too-large population size could make the convergence difficult. Performance is sensitive to algorithm parametrization. Therefore, it is recommended for novices in optimization to employ an algorithm that possesses some parametrization guidelines or select an algorithm with few parameters to adjust. Nonetheless, most metaheuristics available in Matlab have "default" parametrization values. In [13], it has been argued that using "default" values is a rational and justified choice, considering parametrization tuning might be an expensive and lengthy process.

Another common element to any metaheuristics is the user-defined objective function(s), i.e., $f_i(x)$ $(i = 1, 2, \ldots, M)$, where M is the number of objective functions. Customized objective functions is one reason metaheuristics are suitable and configurable to a broad range of optimization problems. Besides choosing the optimization algorithm, the control designer has to determine one or many objective functions. The control designer could choose among hundreds of existing performance criteria or, alternatively, may want to create their own. Either way, there is no good or bad objective functions, but more suitable/relevant ones depending on the system at hand. The objective functions must reflect the controller's functional requirements. For example, control systems used to position industrial grinder tables must avoid overshoot, i.e., preventing tables from going beyond the reference position because it is detrimental to the pieces being ground [31]. For this system, an appropriate objective function targets the overshoot. One way to do so is by using one of the four typical measures of controlled system performance. Those criteria are: the integral time absolute error (ITAE), integral square error (ISE), integral time square error (ITSE), and integral absolute error (IAE). A lower criterion value means less overshoot, corresponding to a better solution according to this requirement. Penalizing a feature in the system response with an objective function is called a soft constraint. Ideally, the objective function reaches the zero value, but the system design still works if it is not the case. If functional requirements must be satisfied to make the design acceptable, it is better to design these requirements as constraints in the optimization problem (1.2).

An additional characteristic of metaheuristics is the use of memory to direct the search, either short-term or long-term memory. Not every metaheuristic uses memory; however, research supports that memory is known as a fundamental component of a powerful metaheuristic [33]. The evidence suggests that most population-based metaheuristics employ the concept of memory. For instance, the ACO algorithm has a long-term memory. A pheromone matrix keeps track of the quality of the visited solutions. At each iteration, the algorithm updates its

Fig. 2.2 Metaheuristic
flowchart

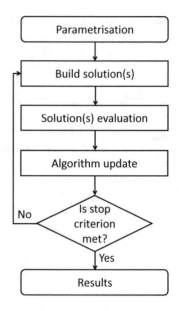

pheromone matrix based on the candidate solutions' quality, and the pheromone matrix guides the building of the next iteration's solutions. Along these lines, different characteristics such as trajectory-based algorithms versus population-based algorithms divide metaheuristics. Trajectory-based methods process a single solution at the time, i.e., the algorithm starts with an initial solution, and the solution is updated by another one at each optimization step. On the opposite, population-based algorithms deal with a set of solutions, in which, at each iteration, new solutions replace some members. Since this classification describes how metaheuristics operate the optimization, the next two sections present a trajectory-based algorithm, SA, and a population-based algorithm, GA. However, although every metaheuristic has a unique way to explore the search space and to select the best solution(s), most metaheuristics follow a general framework, which is presented in Fig. 2.2.

Once the controller tuning problem is modeled as (1.2), the control designer starts the optimization by parameterizing the metaheuristics. After that, the algorithm builds a new solution for trajectory-based algorithms or several new solutions for population-based algorithms. In the controller tuning context, this corresponds to create a set or sets of parameters for the controller(s). The solution(s) is evaluated on the system by the objective function or multiple objective functions. The algorithm update includes selecting the best solution reached so far and performs the steps required to get the algorithm ready for the next iteration, which would be, for example, the update of the pheromone matrix in the ACO algorithm. The algorithm iterates until the stop criterion is reached.

2.4 Simulated Annealing

The simulated annealing algorithm (SA) was originally developed by Kirkpatrick et al. [156]. The heating and cooling of material in metallurgy inspires the SA method. When heated, atoms of some material do not take the strongest structure when they cool naturally, but the most solid structure is reached by controlling the cooling process. By analogy, the SA algorithm attempts to minimize the system's energy, which corresponds to the objective function value by exploring the search space according to artificial system temperature. Figure 2.3 presents the flowchart of SA [160].

The algorithm starts with an initial point and temperature. Usually, the initial point is generated randomly between the bound constraints, and the designer provides the initial temperature. The temperature plays the most critical role in the SA algorithm as it controls the algorithm search. The initial point also influences the algorithm since it is from this point a trial point is generated. At each iteration, the algorithm creates the trial point from the current point based on a probability

Fig. 2.3 Simulated annealing flowchart

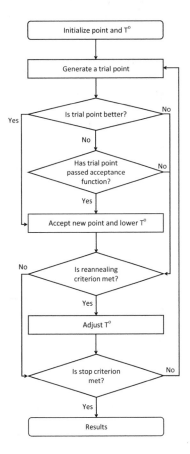

distribution, which latter depends on the current temperature. There are different functions to generate trial points affecting how close or far from the current point the trial point is. After that, the algorithm evaluates the trial point. If the objective function is better than the current, the trial point replaces the current point and becomes the next point. Otherwise, the algorithm accepts the point only if the acceptance function permits. The probability of acceptance is a function of the temperature. Accepting a worse solution may allow the algorithm to reach a new region with a better minimum. Thereafter, the best solution reached so far is stored, and the temperature is reduced. Several functions to update the temperature exist in the literature, such as Boltzmann [123]. The main differences between functions are the speed at which the temperature decreases. After a certain number of iterations, the algorithm reanneals, which makes the temperature rise. The reannealing step prevents the algorithm from being trapped in a region of a local minimum. This action is mostly of help for functions with several local minimums and flat surfaces. The algorithm stops when a stop criterion is reached. Some of the most common stop criteria for SA are as follows [280]:

- *Tolerance on the objective function*: The algorithm stops when the value of the objective function average change is less than a threshold.
- *Iteration numbers*: The algorithm runs until a certain number of iterations is reached.
- *Function evaluation numbers*: The algorithm stops when the maximum number of function evaluations is attained.
- *Maximum time*: The algorithm runs until a specified time is reached.
- *Objective limit*: The algorithm stops when the objective function value reaches a pre-established value.

The SA algorithm used in this monograph is available in MATLAB [280]. Studies of convergence of SA are available in [24, 119, 174].

2.5   Genetics Algorithm

John Holland was the instigator of GA. His inspiration was Darwin's theory of evolution, which loosely refers to the natural evolution of species among a population. The weakest individuals have less chance to survive than the strongest or most adapted ones. For controller tuning, an individual is a controller's parameter set. In an optimization perspective, a population of individuals evolves according to three genetic operations: (*a*) selection, (*b*) crossover, and (*c*) mutation. An individual in the population represents a potential solution, which quality is assessed by an objective function. An individual with a better objective function has a higher chance of survival for the next generation. The initial population is usually created randomly between the bound constraints. The selection operator, also called reproduction, selects the individuals for reproduction based on a probability rule such that the individuals that best performed at the current iteration have more

Fig. 2.4 Genetic algorithm
flowchart

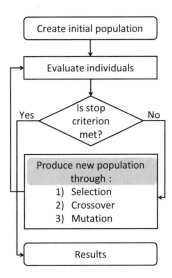

chance to be chosen. Afterward, the crossover operation generates new individuals in the population, combining two selected individuals to create better individuals. Subsequently, the mutation performs random changes on some individuals to possibly find better individuals that do not already exist in the population and cannot be created through the crossover operator. Thereafter, the individuals are evaluated with the objective function. The three operators are performed again until a stop criterion is reached. Figure 2.4 presents the flowchart of GA [125].

Different stop criteria for population-based algorithms exist, follows [88, 330]:

1. *Error-based criteria*: The algorithm stops after reaching a solution within a predetermined range around the optimal solution. Thus, this criterion can only be used for problems in which the optimum is known beforehand.
2. *Exhaustion-based criteria*: The algorithm terminates when a predefined number of iterations or function evaluations is reached or when a pre-established CPU-time is attained.
3. *Improvement-based criteria*: The algorithm stops when the objective function improvement is less than a threshold.
4. *Movement-based criteria*: The algorithm terminates when the population is below a predefined value. The movement can be assessed with the objective function average, meaning the algorithm stops when the average reaches a lower value than the threshold.
5. *Distribution-based criteria*: The algorithm terminates when the population gets concentrated over a small region, hopefully, around the optimum. For instance, the largest distance to the best solution so far is lower than a specified value.
6. *Combined criteria*: It is also possible to combine several stop criteria such that the algorithm stops when one is met.

The designer defines the stop criteria depending on the problem to optimize. The most common practice, when comparing algorithms performance is to employ error-based criteria and/or exhaustion-based criteria.

Along the same line, several ways to perform the selection/reproduction, crossover, and mutation exist. The most used selection operations are the roulette-wheel and the stochastic universal sampling [30]. Some of the most known crossovers are the blend crossover (BLX-α), the simulated binary crossover (SBX), the simplex crossover (SPX), the parent centric crossover (PCX), and the linear crossover [75]. The mutation operator includes actions such as single point, heuristic, and two points [280]. All these operator variants aim to achieve better performance e.g., getting solutions closer to the optimal or getting optimized solutions faster. The designer employing GA for optimization should carefully select the operators. The following publication gives insight into selecting the most suitable operators [172].

The GA algorithm chosen in this monograph is detailed in [30] and retrieved in [29]. Several studies and analyses of convergences for GA are available in the literature [73, 242] and [281] is among the most cited.

References

1. Abbass, H.A.: MBO: marriage in honey bees optimization-A haplometrosis polygynous swarming approach. In: Proceedings of the 2001 Congress on Evolutionary Computation (IEEE Cat. No. 01TH8546), vol. 1, pp. 207–214. IEEE, Piscataway, May 2001
2. Abdechiri, M., Meybodi, M.R., Bahrami, H.: Gases Brownian motion optimization: an algorithm for optimization (GBMO). Appl. Soft Comput. **13**(5), 2932–2946 (2013)
3. Abedinia, O., Amjady, N., Ghasemi, A.: A new metaheuristic algorithm based on shark smell optimization. Complexity **21**(5), 97–116 (2016)
4. Abedinpourshotorban, H., Shamsuddin, S.M., Beheshti, Z., Jawawi, D.N.: Electromagnetic field optimization: a physics-inspired metaheuristic optimization algorithm. Swarm Evol. Comput. **26**, 8–22 (2016)
5. Adham, M.T., Bentley, P.J.: An artificial ecosystem algorithm applied to static and dynamic travelling salesman problems. In: 2014 IEEE International Conference on Evolvable Systems. IEEE, Piscataway (2014)
6. Ahrari, A., Atai, A.A.: Grenade explosion method—a novel tool for optimization of multi-modal functions. Appl. Soft Comput. **10**(4), 1132–1140 (2010)
7. Al-Azza, A.A., Al-Jodah, A.A., Harackiewicz, F.J.: Spider monkey optimization: a novel technique for antenna optimization. IEEE Antennas Wirel. Propag. Lett. **15**, 1016–1019 (2016)
8. Alatas, B.: ACROA: artificial chemical reaction optimization algorithm for global optimization. Expert Syst. Appl. **38**(10), 13170–13180 (2011)
9. Alejo Machado, O.J., Fernández-Luna, J.M., Huete, J.F., Morales, E.R.C.: Fisherman search procedure. Prog. Artif. Intell. **2**(4), 193–203 (2014)
10. Altun, M., Pekcan, O.: A modified approach to cross entropy method: Elitist stepped distribution algorithm. Appl. Soft Comput. **58**, 756–769 (2017)
11. An, J., Kang, Q., Wang, L., Wu, Q.: Mussels wandering optimization: an ecologically inspired algorithm for global optimization. Cognit. Comput. **5**(2), 188–199 (2013)

12. Angeline, P.J., Saunders, G.M., Pollack, J.B.: An evolutionary algorithm that constructs recurrent neural networks. IEEE Trans. Neural Netw. **5**(1), 54–65 (1994)
13. Arcuri, A., Fraser, G.: Parameter tuning or default values? An empirical investigation in search-based software engineering. Empir. Softw. Eng. **18**(3), 594–623 (2013)
14. Arora, S., Singh, S.: Butterfly optimization algorithm: a novel approach for global optimization. Soft Comput. **23**(3), 715–734 (2019)
15. Ashrafi, S.M., Dariane, A.B.: Performance evaluation of an improved harmony search algorithm for numerical optimization: Melody Search (MS). Eng. Appl. Artif. Intell. **26**(4), 1301–1321 (2013)
16. Askarzadeh, A.: Bird mating optimizer: an optimization algorithm inspired by bird mating strategies. Commun. Nonlinear Sci. Numer. Simul. **19**(4), 1213–1228 (2014)
17. Askarzadeh, A.: A novel metaheuristic method for solving constrained engineering optimization problems: crow search algorithm. Comput. Struct. **169**, 1–12 (2016)
18. Atashpaz-Gargari, E., Lucas, C.: Imperialist competitive algorithm: an algorithm for optimization inspired by imperialistic competition. In: IEEE Congress on Evolutionary Computation, pp. 4661–4667 (2007)
19. Babayan, N., Tahani, M.: Team Arrangement Heuristic Algorithm (TAHA): theory and application. Math. Comput. Simul. **166**, 155–176 (2019)
20. Bandieramonte, M., Di Stefano, A., Morana, G.: Grid jobs scheduling: the alienated ant algorithm solution. Multiagent Grid Syst. **6**(3), 225–243 (2010)
21. Barresi, K.M.: Foraging agent swarm optimization with applications in data clustering. In: International Conference on Swarm Intelligence, pp. 230–237. Springer, Berlin (2014)
22. Baykasoglu, A., Akpinar, Ş.: Weighted Superposition Attraction (WSA): a swarm intelligence algorithm for optimization problems–Part 1: unconstrained optimization. Appl. Soft Comput. **56**, 520–540 (2017)
23. Bayraktar, Z., Komurcu, M., Werner, D.H.: Wind Driven Optimization (WDO): a novel nature-inspired optimization algorithm and its application to electromagnetics. In: 2010 IEEE Antennas and Propagation Society International Symposium, pp. 1–4 (2010)
24. Bertsimas, D., Tsitsiklis, J.: Simulated annealing. Stat. Sci. **8**(1), 10–15 (1993)
25. Beskirli, M., Koc, I., Haklı, H., Kodaz, H.: A new optimization algorithm for solving wind turbine placement problem: binary artificial algae algorithm. Renew. Energy **121**, 301–308 (2018)
26. Bidar, M., Mouhoub, M., Sadaoui, S.: Discrete firefly algorithm: a new metaheuristic approach for solving constraint satisfaction problems. In: 2018 IEEE Congress on Evolutionary Computation (CEC), pp. 1–8. IEEE, Piscataway, July 2018
27. Biyanto, T.R., Fibrianto, H.Y., Nugroho, G., Hatta, A.M., Listijorini, E., Budiati, T., Huda, H.: Duelist algorithm: an algorithm inspired by how duelist improve their capabilities in a duel. In: International Conference on Swarm Intelligence, pp. 39–47. Springer, Cham, June 2016
28. Biyanto, T.R., Irawan, S., Febrianto, H.Y., Afdanny, N., Rahman, A.H., Gunawan, K.S., Bethiana, T.N.: Killer whale algorithm: an algorithm inspired by the life of killer whale. Procedia Comput. Sci. **124**, 151–157 (2017)
29. Blasco, X.: Basic genetic algorithm (2020). https://www.mathworks.com/matlabcentral/fileexchange/39021-basic-genetic-algorithm, MATLAB Central File Exchange. Retrieved 13 Feb 2020
30. Blasco Ferragud, F.X.: Control predictivo basado en modelos mediante técnicas de optimización heurística. Aplicación a procesos no lineales y multivariables [Tesis doctoral no publicada]. Universitat Politècnica de València (1999). https://doi.org/10.4995/Thesis/10251/15995
31. Blondin, M.J., Sicard, P.: ACO based controller and anti-windup tuning for motion systems with flexible transmission. In: 2013 26th IEEE Canadian Conference on Electrical and Computer Engineering (CCECE), pp. 1–4. IEEE, Piscataway, May 2013
32. Blondin, M.J., Sanchis, J., Sicard, P., Herrero, J.M.: New optimal controller tuning method for an AVR system using a simplified Ant Colony Optimization with a new constrained Nelder–Mead algorithm. Appl. Soft Comput. **62**, 216–229 (2018)

33. Blum, C., Roli, A.: Metaheuristics in combinatorial optimization: overview and conceptual comparison. ACM Comput. Surv. **35**(3), 268–308 (2003)
34. Blum, C. et al.: Hybrid metaheuristics in combinatorial optimization: a survey. Appl. Soft Comput. **11**(6), 4135–4151 (2011)
35. Boettcher, S., Percus, A.: Nature's way of optimizing. Artif. Intell. **119**(1–2), 275–286 (2000)
36. Borji, A., Hamidi, M.: A new approach to global optimization motivated by parliamentary political competitions. Int. J. Innov. Comp. Inf. Control **5**(6), 1643–1653 (2009)
37. Bouhmala, N.: A variable depth search algorithm for binary constraint satisfaction problems. Math. Probl. Eng. **2015** (2015). https://doi.org/10.1155/2015/637809
38. Bozorgi, S.M., Yazdani, S.: IWOA: an improved whale optimization algorithm for optimization problems. J. Comput. Des. Eng. **6**(3), 243–259 (2019)
39. Brabazon, A., Cui, W., O'Neill, M.: The raven roosting optimisation algorithm. Soft Comput. **20**(2), 525–545 (2016)
40. Canayaz, M., Karci, A.: Cricket behaviour-based evolutionary computation technique in solving engineering optimization problems. Appl. Intell. **44**(2), 362–376 (2016)
41. Caraveo, C., Valdez, F., Castillo, O.: A new optimization meta-heuristic algorithm based on self-defense mechanism of the plants with three reproduction operators. Soft Comput. **22**(15), 4907–4920 (2018)
42. Chang, W.L., Kanesan, J., Kulkarni, A.J.: Seed disperser ant algorithm: an evolutionary approach for optimization. In: European Conference on the Applications of Evolutionary Computation, pp. 643–654. Springer, Cham, April 2015
43. Chen, C.C., Tsai, Y.C., Liu, I.I., Lai, C.C., Yeh, Y.T., Kuo, S.Y., Chou, Y.H.: A novel meta-heuristic: Jaguar algorithm with learning behavior. In: 2015 IEEE International Conference on Systems, Man, and Cybernetics, pp. 1595–1600 (2015)
44. Chen, J., Cai, H., Wang, W.: A new metaheuristic algorithm: car tracking optimization algorithm. Soft Comput. **22**(12), 3857–3878 (2018)
45. Cheng, M.Y., Prayogo, D.: Symbiotic organisms search: a new metaheuristic optimization algorithm. Comput. Struct. **139**, 98–112 (2014)
46. Chetty, S., Adewumi, A.O.: A study on the enhanced best performance algorithm for the just-in-time scheduling problem. Discrete Dyn. Nat. Soc. **2015** (2015). http://dx.doi.org/10.1155/2015/350308
47. Chu, S.C., Tsai, P.W., Pan, J.S.: Cat swarm optimization. In: Pacific Rim International Conference on Artificial Intelligence, pp. 854–858. Springer, Berlin (2006)
48. Civicioglu, P.: Transforming geocentric Cartesian coordinates to geodetic coordinates by using differential search algorithm. Comput. Geosci. **46**, 229–247 (2012)
49. Civicioglu, P.: Artificial cooperative search algorithm for numerical optimization problems. Inf. Sci. **229**, 58–76 (2013)
50. Civicioglu, P.: Backtracking search optimization algorithm for numerical optimization problems. Appl. Math. Comput. **219**(15), 8121–8144 (2013)
51. Clements, D.P., Joslin, D.E.: Squeaky wheel optimization. J. Artif. Intell. Res. **10**, 353–373 (1999)
52. Coello, C.C., Lechuga, M.S.: MOPSO: a proposal for multiple objective particle swarm optimization. In: Proceedings of the 2002 Congress on Evolutionary Computation, vol. 2, pp. 1051–1056 (2002)
53. Coelho, F.C., da Silva Junior, I.C., Dias, B.H., Peres, W.B.: Optimal distributed generation allocation using a new metaheuristic. J. Control Autom. Electr. Syst. **29**(1), 91–98 (2018)
54. Cortés-Toro, E., Crawford, B., Gómez-Pulido, J., Soto, R., Lanza-Gutiérrez, J.: A new metaheuristic inspired by the vapour-liquid equilibrium for continuous optimization. Appl. Sci. **8**(11), 2080 (2018)
55. Cuevas, E., Echavarría, A., Ramírez-Ortegón, M.A.: An optimization algorithm inspired by the States of Matter that improves the balance between exploration and exploitation. Appl. Intell. **40**(2), 256–272 (2014)
56. Cuevas, E., Gonzàlez, A., Zaldvár, D., Pérez-Cisneros, M.: An optimisation algorithm based on the behaviour of locust swarms. Int. J. Bio-Inspired Comput. **7**(6), 402–407 (2015)

57. Cutello, V., Narzisi, G., Nicosia, G., Pavone, M.: Clonal selection algorithms: a comparative case study using effective mutation potentials. In: International Conference on Artificial Immune Systems, pp. 13–28. Springer, Berlin (2005)
58. Dai, C., Zhu, Y., Chen, W.: Seeker optimization algorithm. In: International Conference on Computational and Information, pp. 167–176. Springer, Berlin (2006)
59. Dash, T., Sahu, P.K.: Gradient gravitational search: an efficient metaheuristic algorithm for global optimization. J. Comput. Chem. **36**(14), 1060–1068 (2015)
60. De Boer, P.T., Kroese, D.P., Mannor, S., Rubinstein, R.Y.: A tutorial on the cross-entropy method. Ann. Oper. Res. **134**(1), 19–67 (2005)
61. De Jong, K.: Analysis of the behaviour of a class of genetic adaptive systems. PhD thesis, University of Michigan, Ann Arbor (1975)
62. De Melo, V.V.: Kaizen programming. In: Proceedings of the 2014 Annual Conference on Genetic and Evolutionary Computation (2014)
63. Deb, K., Agrawal, S., Pratap, A., Meyarivan, T.: A fast elitist non-dominated sorting genetic algorithm for multi-objective optimization: NSGA-II. In: International Conference on Parallel Problem Solving from Nature, pp. 849–858. Springer, Berlin (2000)
64. Deb, S., Fong, S., Tian, Z.: Elephant search algorithm for optimization problems. In: 2015 Tenth International Conference on Digital Information Management, pp. 249–255 (2015)
65. Dogan, B., Olmez, T.: A new metaheuristic for numerical function optimization: vortex search algorithm. Inf. Sci. **293**, 125–45 (2015)
66. Dorigo, M., Gambardella, L.M.: Ant colony system: a cooperative learning approach to the traveling salesman problem. IEEE Trans. Evol. Comput. **1**(1), 53–66 (1997)
67. Dorigo, M., Stützle, T.: Ant Colony Optimization. MIT Press, Cambridge (2004)
68. Dorigo, M., Maniezzo, V., Colorni, A.: Ant system: optimization by a colony of cooperating agents. IEEE Trans. Syst. Man Cybern. B Cybern. **26**(1), 29–41 (1996)
69. Duan, H., Qiao, P.: Pigeon-inspired optimization: a new swarm intelligence optimizer for air robot path planning. Int. J. Intell. Comput. Cybern. **7**(1), 24–37 (2014)
70. Duan, Q.Y., Gupta, V.K., Sorooshian, S.: Shuffled complex evolution approach for effective and efficient global minimization. J. Optim. Theory Appl. **76**(3), 501–521 (1993)
71. Duman, E., Uysal, M., Alkaya, A.F.: Migrating birds optimization: a new metaheuristic approach and its performance on quadratic assignment problem. Inf. Sci. **217**, 65–77 (2012)
72. Eesa, A.S., Brifcani, A.M.A., Orman, Z.: Cuttlefish algorithm-a novel bio-inspired optimization algorithm. Int. J. Scientific Eng. Res. **4**(9), 1978–1986 (2013)
73. Eiben, A.E., Aarts, E.H., Van Hee, K.M.: Global convergence of genetic algorithms: a Markov chain analysis. In: International Conference on Parallel Problem Solving from Nature, pp. 3–12. Springer, Berlin, Oct 1990
74. El-Dosuky, M., El-Bassiouny, A., Hamza, T., Rashad, M.: New hoopoe heuristic optimization (2012). Preprint, arXiv:1211.6410
75. Elsayed, S.M., Sarker, R.A., Essam, D.L.: A comparative study of different variants of genetic algorithms for constrained optimization. In: Deb, K. et al. (eds) Simulated Evolution and Learning, SEAL 2010. Lecture Notes in Computer Science, vol. 6457. Springer, Berlin (2010)
76. Emami, H., Derakhshan, F.: Election algorithm: a new socio-politically inspired strategy. AI Commun. **28**(3), 591–603 (2015)
77. Erol, O.K., Eksin, I.: A new optimization method: big bang–big crunch. Adv. Eng. Softw. **37**(2), 106–111 (2006)
78. Eskandar, H., Sadollah, A., Bahreininejad, A., Hamdi, M.: Water cycle algorithm-a novel metaheuristic optimization method for solving constrained engineering optimization problems. Comput. Struct. **110**, 151–166 (2012)
79. Eusuff, M., Lansey, K., Pasha, F.: Shuffled frog-leaping algorithm: a memetic meta-heuristic for discrete optimization. Eng. Optim. **38**(2), 129–154 (2006)
80. Faisal, M., Mathkour, H., Alsulaiman, M.: AntStar: enhancing optimization problems by integrating an Ant System and A* algorithm. Sci. Program. **2016**, 1–12 (2016)

81. Faramarzi, A., Heidarinejad, M., Stephens, B., Mirjalili, S.: Equilibrium optimizer: a novel optimization algorithm. Knowl. Based Syst. (2019). https://doi.org/10.1016/j.knosys.2019.105190

82. Farasat, A., Menhaj, M.B., Mansouri, T., Moghadam, M.R.S.: ARO: a new model-free optimization algorithm inspired from asexual reproduction. Appl. Soft Comput. **10**(4), 1284–1292 (2010)

83. Farmer, J.D., Packard, N.H., Perelson, A.S.: The immune system, adaptation, and machine learning. Physica D **22**(1–3), 187–204 (1986)

84. Feng, X., Lau, F.C., Yu, H.: A novel bio-inspired approach based on the behavior of mosquitoes. Inf. Sci. **233**, 87–108 (2013)

85. Feng, X., Zou, R., Yu, H.: A novel optimization algorithm inspired by the creative thinking process. Soft Comput. **19**(10), 2955–2972 (2015)

86. Feng, X., Ma, M., Yu, H.: Crystal energy optimization algorithm. Comput. Intell. **32**(2), 284–322 (2016)

87. Feo, T.A., Resende, M.G.: A probabilistic heuristic for a computationally difficult set covering problem. Oper. Res. Lett. **8**(2), 67–71 (1989)

88. Fernández-Vargas, J.A., Bonilla-Petriciolet, A., Rangaiah, G.P., Fateen, S.E.K.: Performance analysis of stopping criteria of population-based metaheuristics for global optimization in phase equilibrium calculations and modeling. Fluid Phase Equilib. **427**, 104–125 (2016)

89. Filipović, V., Kartelj, A., Matić, D.: An electromagnetism metaheuristic for solving the maximum betweenness problem. Appl. Soft Comput. **13**(2), 1303–1313 (2013)

90. Findik, O.: Bull optimization algorithm based on genetic operators for continuous optimization problems. Turk. J. Electr. Eng. Comput. Sci. **23**(Suppl 1), 2225–2239 (2015)

91. Floudas, C.A., Pardalos, P.M.: A Collection of Test Problems for Constrained Global Optimization Algorithms. Springer Science & Business Media, Berlin, 15 Sept 1990

92. Fogel, L.G., Owens, A.J., Walsh, M.J.: Artificial Intelligence Through Simulated Evolution. Wiley, New York (1966)

93. Fulkerson, D.R., Nemhouser, G.L., Trotter, L.E. Jr.: Two computationally difficult set covering problems that arise in computing the 1-width of incidence matrices of Steiner triple systems. Cornell Univ Ithaca NY Dept of Operations Research, Nov 1973

94. Gandomi, A.H.: Interior search algorithm (ISA): a novel approach for global optimization. ISA Trans. **53**(4), 1168–1183 (2014)

95. Gandomi, A.H., Alavi, A.H.: Krill herd: a new bio-inspired optimization algorithm. Commun. Nonlinear Sci. Numer. Simul. **17**(12), 4831–4845 (2012)

96. Gang, M., Wei, Z., Xiaolin, C.: A novel particle swarm optimization algorithm based on particle migration. Appl. Math. Comput. **218**(11), 6620–6626 (2012)

97. Gao-Wei, Y., Zhanju, H.: A novel atmosphere clouds model optimization algorithm, In: 2012 International Conference on Computing, Measurement, Control and Sensor Network, Taiyuan, pp. 217–220 (2012). https://doi.org/10.1109/CMCSN.2012.117

98. Gargari, E.A., Hashemzadeh, F., Rajabioun, R., Lucas, C.: Colonial competitive algorithm. Int. J. Intell. Comput. Cybern. **1**(3), 337–355 (2008)

99. Geem, Z.W., Kim, J.H., Loganathan, G.V.: A new heuristic optimization algorithm: harmony search. Simulation **76**(2), 60–68 (2001)

100. Ghaemi, M., Feizi-Derakhshi, M.R.: Forest optimization algorithm. Expert Syst. Appl. **41**(15), 6676–6687 (2014)

101. Gheraibia, Y., Moussaoui, A.: Penguins search optimization algorithm (PeSOA). In: International Conference on Industrial, Engineering and Other Applications of Applied Intelligent Systems. Springer, Berlin (2013)

102. Gholizadeh, S., Ebadijalal, M.: Performance based discrete topology optimization of steel braced frames by a new metaheuristic. Adv. Eng. Softw. **123**, 77–92 (2018)

103. Ghorbani, N., Babaei, E.: Exchange market algorithm. Appl. Soft Comput. **19**, 177–187 (2014)

104. Glover, F.: Tabu search—part I. ORSA J. Comput. **1**(3), 190–206 (1989)

105. Glover, F., Laguna, M., Martí, R.: Scatter search. In: Advances in Evolutionary Computing, pp. 519–537. Springer, Berlin (2003)
106. Goncalves, R., Goldbarg, M.C., Goldbarg, E.F., Delgado, M.R.: Warping search: a new metaheuristic applied to the protein structure prediction. In: Proceedings of the 10th Annual Conference on Genetic and Evolutionary Computation, pp. 349–350 (2008)
107. Goncalves, M.S., Lopez, R.H., Miguel, L.F.: Search group algorithm: a new metaheuristic method for the optimization of truss structures. Comput. Struct. **153**, 165–84 (2015)
108. Gonzalez-Fernandez, Y., Chen, S.: Leaders and followers—a new metaheuristic to avoid the bias of accumulated information. In: 2015 IEEE Congress on Evolutionary Computation, pp. 776–783 (2015)
109. Haddad, O.B., Afshar, A., Mariño, M.A.: Honey-bees mating optimization (HBMO) algorithm: a new heuristic approach for water resources optimization. Water Resour. Manag. **20**(5), 661–680 (2006)
110. Hajipour, H., Khormuji, H.B., Rostami, H.: ODMA: a novel swarm-evolutionary metaheuristic optimizer inspired by open source development model and communities. Soft Comput. **20**(2), 727–747 (2016)
111. Haldar, V., Chakraborty, N.: A novel evolutionary technique based on electrolocation principle of elephant nose fish and shark: fish electrolocation optimization. Soft Comput. **21**(14), 3827–3848 (2017)
112. Harifi, S., Khalilian, M., Mohammadzadeh, J., Ebrahimnejad, S.: Emperor penguins colony: a new metaheuristic algorithm for optimization. Evol. Intell. **12**(2), 211–226 (2019)
113. Hasancebi, O., Azad, S.K.: Adaptive dimensional search: a new metaheuristic algorithm for discrete truss sizing optimization. Comput. Struct. **154**, 1–6 (2015)
114. Hassani, K., Lee, W.S.: Multi-objective design of state feedback controllers using reinforced quantum-behaved particle swarm optimization. Appl. Soft Comput. **41**, 66–76 (2016)
115. Hatamlou, A.: Black hole: a new heuristic optimization approach for data clustering. Inf. Sci. **222**, 175–184 (2013)
116. Hatamlou, A.: Heart: a novel optimization algorithm for cluster analysis. Prog. Artif. Intell. **2**(2–3), 167–173 (2014)
117. He, S., Wu, Q.H., Saunders, J.R.: Group search optimizer: an optimization algorithm inspired by animal searching behavior. IEEE Trans. Evol. Comput. **13**(5), 973–990 (2009)
118. Heidari, A.A., Mirjalili, S., Faris, H., Aljarah, I., Mafarja, M., Chen, H.: Harris hawks optimization: algorithm and applications. Future Gener. Comput. Syst. **97**, 849–872 (2019)
119. Henderson, D., Jacobson, S.H., Johnson, A.W.: The theory and practice of simulated annealing. In: Handbook of Metaheuristics, pp. 287–319. Springer, Boston (2003)
120. Hernandez, H., Blum, C.: Distributed graph coloring: an approach based on the calling behavior of Japanese tree frogs. Swarm Intell. **6**(2), 117–150 (2012)
121. Hock, V., Schittkowski, K.: Test Examples for Nonlinear Programming Codes. Lecture Notes in Economics and Mathematical Systems. Springer, Berlin (1981)
122. Ibrahim, M.K., Ali, R.S.: Novel optimization algorithm inspired by camel traveling behavior. Iraqi J. Electr. Electron. Eng. **12**(2), 167–177 (2016)
123. Ingber, L.: Adaptive simulated annealing (ASA): lessons learned (2000). Preprint, arXiv cs/0001018
124. Iordache, S.: Consultant-guided search: a new metaheuristic for combinatorial optimization problems. In: Proceedings of the 12th Annual Conference on Genetic and Evolutionary Computation, pp. 225–232 (2010)
125. Jaen-Cuellar, A.Y., de J. Romero-Troncoso, R., Morales-Velazquez, L., Osornio-Rios, R.A.: PID-controller tuning optimization with genetic algorithms in servo systems. Int. J. Adv. Robot. Syst. **10**(9), 324 (2013)
126. Jahani, E., Chizari, M.: Tackling global optimization problems with a novel algorithm–Mouth Brooding Fish algorithm. Appl. Soft Comput. **62**, 987–1002 (2018)
127. Jain, M., Maurya, S., Rani, A., Singh, V.: Owl search algorithm: a novel nature-inspired heuristic paradigm for global optimization. J. Intell. Fuzzy Syst. **34**(3), 1573–1582 (2018)

128. Jain, M., Singh, V., Rani, A.: A novel nature-inspired algorithm for optimization: squirrel search algorithm. Swarm Evol. Comput. **44**, 148–175 (2019)
129. James, J.Q., Li, V.O.: A social spider algorithm for global optimization. Appl. Soft Comput. **30**, 614–627 (2015)
130. Javidy, B., Hatamlou, A., Mirjalili, S.: Ions motion algorithm for solving optimization problems. Appl. Soft Comput. **32**, 72–79 (2015)
131. Jin, X., Reynolds, R.G.: Using knowledge-based evolutionary computation to solve nonlinear constraint optimization problems: a cultural algorithm approach. In: Proceedings of the 1999 Congress on Evolutionary Computation-CEC99 (Cat. No. 99TH8406), vol. 3, pp. 1672–1678. IEEE, Piscataway, July 1999
132. Kaboli, S.H., Selvaraj, J., Rahim, N.A.: Rain-fall optimization algorithm: a population based algorithm for solving constrained optimization problems. J. Comput. Sci. **19**, 31–42 (2017)
133. Kadioglu, S., Sellmann, M.: Dialectic search. In: International Conference on Principles and Practice of Constraint Programming. Springer, Berlin (2009)
134. Kallioras, N.A., Lagaros, N.D., Avtzis, D.N.: Pity beetle algorithm–a new metaheuristic inspired by the behavior of bark beetles. Adv. Eng. Softw. **121**, 147–166 (2018)
135. Karaboga, D., Basturk, B.: Artificial bee colony (ABC) optimization algorithm for solving constrained optimization problems. In: International Fuzzy Systems Association World Congress, pp. 789–798, Springer, Berlin (2007)
136. Karami, H., Sanjari, M.J., Gharehpetian, G.B.: Hyper-Spherical Search (HSS) algorithm: a novel meta-heuristic algorithm to optimize nonlinear functions. Neural Comput. Appl. **25**(6), 1455–1465 (2014)
137. Karimi, A., Siarry, P.: Global simplex optimization—a simple and efficient metaheuristic for continuous optimization. Eng. Appl. Artif. Intell. **25**(1), 48–55 (2012)
138. Kashan, A.H.: League championship algorithm: a new algorithm for numerical function optimization. In: International Conference of Soft Computing and Pattern Recognition, pp. 43–48 (2009)
139. Kashan, A.H.: A new metaheuristic for optimization: optics inspired optimization (OIO). Comput. Oper. Res. **55**, 99–125 (2015)
140. Kashan, A.H., Tavakkoli-Moghaddam, R., Gen, M.: Find-Fix-Finish-Exploit-Analyze (F3EA) meta-heuristic algorithm: an effective algorithm with new evolutionary operators for global optimization. Comput. Ind. Eng. **128**, 192–218 (2019)
141. Kaveh, A., Bakhshpoori, T.: A new metaheuristic for continuous structural optimization: water evaporation optimization. Struct. Multidiscip. Optim. **54**(1), 23–43 (2016)
142. Kaveh, A., Dadras, A.: A novel meta-heuristic optimization algorithm: thermal exchange optimization. Adv. Eng. Softw. **110**, 69–84 (2017)
143. Kaveh, A., Farhoudi, N.: A new optimization method: dolphin echolocation. Adv. Eng. Softw. **59**, 53–70 (2013)
144. Kaveh, A., Ghazaan, M.I.: A new meta-heuristic algorithm: vibrating particles system. Sci. Iran. Trans. A Civil Eng. **24**(2), 551 (2017)
145. Kaveh, A., Khayatazad, M.: A new meta-heuristic method: ray optimization. Comput. Struct. **112**, 283–294 (2012)
146. Kaveh, A., Mahdavi, V.R.: Colliding bodies optimization: a novel meta-heuristic method. Comput. Struct. **139**, 18–27 (2014)
147. Kaveh, A., Talatahari, S.: A novel heuristic optimization method: charged system search. Acta Mech. **213**(3–4), 267–289 (2010)
148. Kaveh, A., Zolghadr, A.: A novel meta-heuristic algorithm: tug of war optimization. Iran Univ. Sci. Technol. **6**(4), 469–492 (2016)
149. Kaveh, A., Zolghadr, A.: Cyclical parthenogenesis algorithm: a new meta-heuristic algorithm. Asian J. Civil Eng. **18**(5), 673–701 (2017)
150. Kaveh, A., Share, M.A., Moslehi, M.: Magnetic charged system search: a new meta-heuristic algorithm for optimization. Acta Mech. **224**(1), 85–107 (2013)

151. Kaveh, A., Mirzaei, B., Jafarvand, A.: An improved magnetic charged system search for optimization of truss structures with continuous and discrete variables. Appl. Soft Comput. **28**, 400–410 (2015)
152. Kawabe, T., Tagami, T.: A real coded genetic algorithm for matrix inequality design approach of robust PID controller with two degrees of freedom. In: Proceedings of 12th IEEE International Symposium on Intelligent Control, Istanbul, pp. 119–124 (1997)
153. Kennedy, J., Eberhart, R.: Particle swarm optimization (PSO). In: Proc. IEEE International Conference on Neural Networks, Perth (1995)
154. Khatri, A., Gaba, A., Rana, K.P., Kumar, V.: A novel life choice-based optimizer. Soft Comput. **6**, 1–21 (2019)
155. Kiran, M.S.: TSA: tree-seed algorithm for continuous optimization. Expert Syst. Appl. **42**(19), 6686–6698 (2015)
156. Kirkpatrick, S., Gelatt, C.D., Vecchi, M.P.: Optimization by simulated annealing. Science **220**(4598), 671–680 (1983)
157. Kiruthiga, G., Krishnapriya, S., Karpagambigai, V., Pazhaniraja, N., Paul, P.V.: A novel Bio-inspired algorithm based on the foraging behaviour of the Bottlenose dolphin. In: International Conference on Computation of Power, Energy, Information and Communication (ICCPEIC) (2015)
158. Krishnanand, K.N., Ghose, D.: Glowworm swarm optimisation: a new method for optimising multi-modal functions. Int. J. Comput. Intell. Stud. **1**(1), 93–119 (2009)
159. Kuo, R.J., Zulvia, F.E.: The gradient evolution algorithm: a new metaheuristic. Inf. Sci. **316**, 246–265 (2015)
160. Lalaoui, M., El Afia, A., Chiheb, R.: Simulated annealing with adaptive neighborhood using fuzzy logic controller. In: Proceedings of LOPAL '18 (2018)
161. Lam, A.Y.S., Li, V.O.K.: Chemical-reaction-inspired metaheuristic for optimization. IEEE Trans. Evol. Comput. **14**(3), 381–399 (2009)
162. Lamy, J.B.: Artificial Feeding Birds (AFB): a new metaheuristic inspired by the behavior of pigeons. In: Advances in Nature-Inspired Computing and Applications, pp. 43–60. Springer, Cham (2019)
163. Lee, J.W., Lee, J.Y., Lee, J.J.: Jenga-inspired optimization algorithm for energy-efficient coverage of unstructured WSNs. IEEE Wirel. Commun. Lett. **2**(1), 34–37 (2012)
164. Li, X.L.: An optimizing method based on autonomous animats: fish-swarm algorithm. Syst. Eng. Theory Pract. **22**(11), 32–38 (2002)
165. Li, X.L.: A new intelligent optimization-artificial fish swarm algorithm. PhD thesis, Zhejiang University, China, June 2003
166. Li, X., Zhang, J., Yin, M.: Animal migration optimization: an optimization algorithm inspired by animal migration behavior. Neural Comput. Appl. **24**(7–8), 1867–1877 (2014)
167. Li, W., Wang, L., Yao, Q., Jiang, Q., Yu, L., Wang, B., Hei, X.: Cloud particles differential evolution algorithm: a novel optimization method for global numerical optimization. Math. Probl. Eng. **2015**, 1–36 (2015)
168. Li, M.D., Zhao, H., Weng, X.W., Han, T.: A novel nature-inspired algorithm for optimization: virus colony search. Adv. Eng. Softw. **92**, 65–88 (2016)
169. Liang, Y.-C., Cuevas Juarez, J.R.: A novel metaheuristic for continuous optimization problems: virus optimization algorithm. Eng. Optim. **48**(1), 73–93 (2016)
170. Liang, J.J., Qin, A.K., Suganthan, P.N., Baskar, S.: Comprehensive learning particle swarm optimizer for global optimization of multimodal functions. IEEE Trans. Evol. Comput. **10**(3), 281–295 (2006)
171. Liang, X., Li, W., Liu, P., Zhang, Y., Agbo, A.A.: Social network-based swarm optimization algorithm. In: 2015 IEEE 12th International Conference on Networking, Sensing and Control, pp. 360–365 (2015)
172. Lim, S.M., Sultan, A.B.M., Sulaiman, M.N., Mustapha, A., Leong, K.Y.: Crossover and mutation operators of genetic algorithms. Int. J. Mach. Learn. Comput. **7**(1), 9–12 (2017)
173. Liu, Y., Tian, P.: A multi-start central force optimization for global optimization. Appl. Soft Comput. **27**, 92–98 (2015)

174. Locatelli, M.: Simulated annealing algorithms for continuous global optimization: convergence conditions. J. Optim. Theory Appl. **104**(1), 121–133 (2000)
175. Luo, F., Zhao, J., Dong, Z.Y.: A new metaheuristic algorithm for real-parameter optimization: natural aggregation algorithm. In: IEEE Congress on Evolutionary Computation (CEC), pp. 94–103 (2016)
176. Ma, L., Zhu, Y., Liu, Y., Tian, L., Chen, H.: A novel bionic algorithm inspired by plant root foraging behaviors. Appl. Soft Comput. **37**, 95–113 (2015)
177. Ma, L., Wang, X., Huang, M., Zhang, H., Chen, H.: A novel evolutionary root system growth algorithm for solving multi-objective optimization problems. Appl. Soft Comput. **57**, 379–398 (2017)
178. Marin, J., Sole, R.V.: Macroevolutionary algorithms: a new optimization method on fitness landscapes. IEEE Trans. Evol. Comput. **3**(4), 272–286 (1999)
179. Marinakis, Y., Marinaki, M., Matsatsinis, N.: A bumble bees mating optimization algorithm for global unconstrained optimization problems. In: Nature Inspired Cooperative Strategies for Optimization (NICSO 2010), pp. 305–318. Springer, Berlin (2010)
180. Mehrabian, A.R., Lucas, C.: A novel numerical optimization algorithm inspired from weed colonization. Ecol. Inform. **1**(4), 355–366 (2006)
181. Meignan, D., Koukam, A., Créput, J.C.: Coalition-based metaheuristic: a self-adaptive metaheuristic using reinforcement learning and mimetism. J. Heuristics **16**(6), 859–879 (2010)
182. Meng, X., Liu, Y., Gao, X., Zhang, H.: A new bio-inspired algorithm: chicken swarm optimization. In: International Conference in Swarm Intelligence, pp. 86–94. Springer, Berlin (2014)
183. Meng, X.B., Gao, X.Z., Lu, L., Liu, Y., Zhang, H.: A new bio-inspired optimisation algorithm: Bird Swarm Algorithm. J. Exp. Theor. Artif. Intell. **28**(4), 673–687 (2016)
184. Merrikh-Bayat, F.: A numerical optimization algorithm inspired by the strawberry plant (2014). Preprint, arXiv:1407.7399
185. Merrikh-Bayat, F.: The runner-root algorithm: a metaheuristic for solving unimodal and multimodal optimization problems inspired by runners and roots of plants in nature. Appl. Soft Comput. **33**, 292–303 (2015)
186. Mirjalili, S.: The ant lion optimizer. Adv. Eng. Softw. **83**, 80–98 (2015)
187. Mirjalili, S.: Moth-flame optimization algorithm: a novel nature-inspired heuristic paradigm. Knowl. Based Syst. **89**, 228–249 (2015)
188. Mirjalili, S.: SCA: a sine cosine algorithm for solving optimization problems. Knowl. Based Syst. **96**, 120–133 (2016)
189. Mirjalili S.: Dragonfly algorithm: a new meta-heuristic optimization technique for solving single-objective, discrete, and multi-objective problems. Neural Comput. Appl. **27**(4), 1053–1073 (2017)
190. Mirjalili, S., Hashim, S.Z.M.: BMOA: binary magnetic optimization algorithm. Int. J. Mach. Learn. Comput. **2**(3), 204 (2012)
191. Mirjalili, S., Lewis, A.: The whale optimization algorithm. Adv. Eng. Softw. **1**(95), 51–67 (2016)
192. Mirjalili, S., Sadiq, A.S.: Magnetic optimization algorithm for training multi layer perceptron. In 2011 IEEE 3rd International Conference on Communication Software and Networks, pp. 42–46. IEEE, Piscataway, 2011 May
193. Mirjalili, S., Mirjalili, S.M., Yang, X.S.: Binary bat algorithm. Neural Comput. Appl. **25**(3–4), 663–681 (2014)
194. Mirjalili, S., Mirjalili, S.M., Lewis, A.: Grey wolf optimizer. Adv. Eng. Softw. **69**, 46–61 (2014)
195. Mirjalili, S., Mirjalili, S.M., Hatamlou, A.: Multi-verse optimizer: a nature-inspired algorithm for global optimization. Neural Comput. Appl. **27**(2), 495–513 (2016)
196. Mirjalili, S., Gandomi, A.H., Mirjalili, S.Z., Saremi, S., Faris, H., Mirjalili, S.M.: Salp Swarm Algorithm: a bio-inspired optimizer for engineering design problems. Adv. Eng. Softw. **114**, 163–191 (2017)

197. Mladenović, N., Hansen, P.: Variable neighborhood search. Comput. Oper. Res. **24**(11), 1097–1100 (1997)
198. Mo, H., Xu, L.: Magnetotactic bacteria optimization algorithm for multimodal optimization. In: 2013 IEEE Symposium on Swarm Intelligence (SIS), pp. 240–247. IEEE, Piscataway, April 2013
199. Moein, S., Logeswaran, R.: KGMO: a swarm optimization algorithm based on the kinetic energy of gas molecules. Inf. Sci. **275**, 127–144 (2014)
200. Moghaddam, F.F., Moghaddam, R.F., Cheriet, M.: Curved space optimization: a random search based on general relativity theory (2012). Preprint, arXiv:1208.2214
201. Moghdani, R., Salimifard, K.: Volleyball premier league algorithm. Appl. Soft Comput. **64**, 161–185 (2018)
202. Mohamed, A.K., Mohamed, A.W., Hadi, A.A.: Gaining-sharing knowledge based algorithm for solving optimization problems: a novel nature-inspired algorithm. Int. J. Mach. Learn. Cybern. **11**, 1501–1529 (2019)
203. Mohseni, S., Gholami, R., Zarei, N., Zadeh, A.R.: Competition over resources: a new optimization algorithm based on animals behavioral ecology. In: 2014 International Conference on Intelligent Networking and Collaborative Systems, pp. 311–315. IEEE, Piscataway, Sept 2014
204. Moosavian, N., Roodsari, B.K.: Soccer league competition algorithm: a novel meta-heuristic algorithm for optimal design of water distribution networks. Swarm Evol. Comput. **17**, 14–24 (2014)
205. Mora-Gutiérrez, R.A., Ramírez-Rodríguez, J., Rincón-García, E.A.: An optimization algorithm inspired by musical composition. Artif. Intell. Rev. **41**(3), 301–315 (2014)
206. Mortazavi, A., Togan, V., Nuhoglu, A.: Interactive search algorithm: a new hybrid metaheuristic optimization algorithm. Eng. Appl. Artif. Intell. **71**, 275–292 (2018)
207. Mozaffari, A., Fathi, A., Behzadipour, S.: The great salmon run: a novel bio-inspired algorithm for artificial system design and optimisation. Int. J. Bio-Inspired Comput. **4**(5), 286–301 (2012)
208. Mua'zu, M.B., Salawudeen, A.T., Sikiru, T.H., Abdu, A.I., Mohammad, A.: Weighted artificial fish swarm algorithm with adaptive behaviour based linear controller design for nonlinear inverted pendulum. J. Eng. Res. **20**(1), 1–12 (2015)
209. Mucherino, A., Seref, O.: Monkey search: a novel metaheuristic search for global optimization. AIP Conf. Proc. **953**(1), 162–173 (2007)
210. Muhammad, A.: MOX: a novel global optimization algorithm inspired from Oviposition site selection and egg hatching inhibition in mosquitoes. Appl. Soft Comput. **11**(8), 4614–4625 (2011)
211. Munoz, M.A., Lopez, J.A., Caicedo, E.: An artificial beehive algorithm for continuous optimization. Int. J. Intell. Syst. **24**(11), 1080–1093 (2009)
212. Murase, H., Wadano, A.: Photosynthetic algorithm for machine learning and TSP. IFAC Proc. Vol. **31**(12), 19–24 (1998)
213. Muthiah-Nakarajan, V., Noel, M.M.: Galactic swarm optimization: a new global optimization metaheuristic inspired by galactic motion. Appl. Soft Comput. **38**, 771–787 (2016)
214. Nakawiro, W., Erlich, I., Rueda, J.L.: A novel optimization algorithm for optimal reactive power dispatch: a comparative study. In: 2011 4th International Conference on Electric Utility Deregulation and Restructuring and Power Technologies (DRPT), pp. 1555–1561. IEEE, Piscataway, July 2011
215. Nakrani, S., Tovey, C.: On honey bees and dynamic server allocation in internet hosting centers. Adapt. Behav. **12**(3–4), 223–240 (2004)
216. Nasir, A.N., Tokhi, M.O., Sayidmarie, O., Ismail, R.R.: A novel adaptive spiral dynamic algorithm for global optimization. In: 2013 13th UK Workshop on Computational Intelligence, pp. 334–341 (2013)
217. Nasuto, S., Bishop, M.: Convergence analysis of stochastic diffusion search. Parallel Algorithms Appl. **14**(2), 89–107 (1999)

218. Nematollahi, A.F., Rahiminejad, A., Vahidi, B.: A novel physical based meta-heuristic optimization method known as Lightning Attachment Procedure Optimization. Appl. Soft Comput. **59**, 596–621 (2017)
219. Niu, B., Wang, H.: Bacterial colony optimization. Discrete Dyn. Nat. Soc. (2012). https://doi.org/10.1155/2012/698057
220. Odili, J.B., Kahar, M.N.: African buffalo optimization (ABO): a new meta-heuristic algorithm. J. Adv. Appl. Sci. **3**(03), 101–106 (2016)
221. Oftadeh, R., Mahjoob, M.J., Shariatpanahi, M.: a novel meta-heuristic optimization algorithm inspired by group hunting of animals: hunting search. Comput. Math. Appl. **60**(7), 2087–2098 (2010)
222. Osaba, E., Diaz, F., Onieva, E.: Golden ball: a novel meta-heuristic to solve combinatorial optimization problems based on soccer concepts. Appl. Intell. **41**(1), 145–166 (2014)
223. Pakzad-Moghaddam, S.H., Mina, H., Mostafazadeh, P.: A novel optimization booster algorithm. Comput. Industr. Eng. **136**, 591–613 (2019)
224. Pan, W.T.: A new fruit fly optimization algorithm: taking the financial distress model as an example. Knowl. Based Syst. **26**, 69–74 (2012)
225. Passino, K.M.: Biomimicry of bacterial foraging for distributed optimization and control. IEEE Control Syst. Mag. **22**(3), 52–67 (2002). https://doi.org/10.1109/MCS.2002.1004010
226. Pedersen, M.E., Chipperfield, A.J.: Simplifying particle swarm optimization. Appl. Soft Comput. **10**(2), 618–628 (2010)
227. Pelikan, M., Goldberg, D.E., Cantú-Paz, E.: BOA: the Bayesian optimization algorithm. In: Proceedings of the 1st Annual Conference on Genetic and Evolutionary Computation, vol. 1, pp. 525–532 (1999)
228. Pierezan, J., Coelho, L.D.S.: Coyote optimization algorithm: a new metaheuristic for global optimization problems. In: 2018 IEEE Congress on Evolutionary Computation (CEC), pp. 1–8. IEEE, Piscataway, July 2018
229. Precup, R.E., David, R.C., Petriu, E.M., Preitl, S., Radac, M.B.: Novel adaptive gravitational search algorithm for fuzzy controlled servo systems. IEEE Trans. Industr. Inform. **8**(4), 791–800 (2012)
230. Premaratne, U., Samarabandu, J., Sidhu, T.: A new biologically inspired optimization algorithm. In: 2009 International Conference on Industrial and Information Systems (ICIIS), pp. 279–284 (2009)
231. Puangdownreong, D.: Spiritual search: a novel metaheuristic algorithm for control engineering optimization. Int. Rev. Automat. Contr. **11**(2), 86–97 (2018)
232. Qin, J.: A new optimization algorithm and its application—key cutting algorithm. In: 2009 IEEE International Conference on Grey Systems and Intelligent Services, pp. 1537–1541 (2009)
233. Rabanal, P., Rodríguez, I., Rubio, F.: Using river formation dynamics to design heuristic algorithms. In: International Conference on Unconventional Computation, pp. 163–177. Springer, Berlin, August 2007
234. Rahmani, R., Yusof, R.: A new simple, fast and efficient algorithm for global optimization over continuous search-space problems: radial movement optimization. Appl. Math. Comput. **248**, 287–300 (2014)
235. Rajabioun R.: Cuckoo optimization algorithm. Appl. Soft Comput. **11**(8), 5508–5518 (2011)
236. Rao, R.V., Savsani, V.J., Vakharia, D.P.: Teaching–learning-based optimization: a novel method for constrained mechanical design optimization problems. Comput. Aided Des. **43**(3), 303–315 (2011)
237. Rashedi, E., Nezamabadi-Pour, H., Saryazdi, S.: GSA: a gravitational search algorithm. Inf. Sci. **179**(13), 2232–2248 (2009)
238. Razmjooy, N., Khalilpour, M., Ramezani, M.: A new meta-heuristic optimization algorithm inspired by FIFA world cup competitions: theory and its application in PID designing for AVR system. J. Control Autom. Electr. Syst. **27**(4), 419–440 (2016)
239. Rbouh, I., El Imrani, A.A.: Hurricane-based optimization algorithm. AASRI Procedia **6**, 26–33 (2014)

240. Reddy, K.S., Panwar, L., Panigrahi, B.K., Kumar, R.: Binary whale optimization algorithm: a new metaheuristic approach for profit-based unit commitment problems in competitive electricity markets. Eng. Optim. **51**(3), 369–389 (2019)
241. Rodzin, S.I.: Smart dispatching and metaheuristic swarm flow algorithm. J. Comput. Syst. Sci. Int. **53**(1), 109–115 (2014)
242. Rudolph, G.: Convergence analysis of canonical genetic algorithms. IEEE Trans. Neural Netw. **5**(1), 96–101 (1994)
243. Sadollah, A., Bahreininejad, A., Eskandar, H., Hamdi, M.: Mine blast algorithm: a new population based algorithm for solving constrained engineering optimization problems. Appl. Soft Comput. **13**(5), 2592–2612 (2013)
244. Salcedo-Sanz, S., Del Ser, J., Landa-Torres, I., Gil-López, S., Portilla-Figueras, J.A.: The coral reefs optimization algorithm: a novel metaheuristic for efficiently solving optimization problems. Scientific World J. (2014). https://doi.org/10.1155/2014/739768
245. Salimi, H.: Stochastic fractal search: a powerful metaheuristic algorithm. Knowl. Based Syst. **75**, 1–8 (2015)
246. Saremi, S., Mirjalili, S.M., Mirjalili, S.: Chaotic krill herd optimization algorithm. Procedia Technol. **12**, 180–185 (2014)
247. Saremi, S., Mirjalili, S., Lewis, A.: Grasshopper optimisation algorithm: theory and application. Adv. Eng. Softw. **105**, 30–47 (2017)
248. Savsani, P., Savsani, V.: Passing vehicle search (PVS): a novel metaheuristic algorithm. Appl. Math. Modell. **40**(5–6), 3951–3978 (2016)
249. Savsani, V. et al.: Pareto optimization of a half car passive suspension model using a novel multiobjective heat transfer search algorithm. Modell. Simul. Eng. **2017** (2017). https://doi.org/10.1155/2017/2034907
250. Shadravan, S., Naji, H.R., Bardsiri, V.K.: The Sailfish Optimizer: a novel nature-inspired metaheuristic algorithm for solving constrained engineering optimization problems. Eng. Appl. Artif. Intell. **80**, 20–34 (2019)
251. Shah-Hosseini, H.: Intelligent water drops algorithm: a new optimization method for solving the multiple knapsack problem. Int. J. Intell. Comput. Cybern. **1**(2), 193–212 (2008)
252. Shah-Hosseini, H.: Principal components analysis by the galaxy-based search algorithm: a novel metaheuristic for continuous optimisation. Int. J. Comput. Sci. Eng. **6**(1–2), 132–140 (2011)
253. Sharafi, Y., Khanesar, M.A., Teshnehlab, M.: Discrete binary cat swarm optimization algorithm. In: 2013 3rd IEEE International Conference on Computer, Control and Communication (IC4), pp. 1–6. IEEE, Piscataway, 25 Sept 2013
254. Sharma, T.K., Pant, M.: Shuffled artificial bee colony algorithm. Soft Comput. **21**(20), 6085–6104 (2017)
255. Sharma, M.K., Phonrattanasak, P., Leeprechanon, N.: Improved bees algorithm for dynamic economic dispatch considering prohibited operating zones. In: 2015 IEEE Innovative Smart Grid Technologies-Asia (ISGT ASIA), pp. 1–6 (2015)
256. Sharma, A., Sharma, A., Panigrahi, B.K., Kiran, D., Kumar, R.: Ageist spider monkey optimization algorithm. Swarm Evol. Comput. **28**, 58–77 (2016)
257. Sharma, H., Bansal, J.C., Arya, K.V., Yang, X.S.: Lévy flight artificial bee colony algorithm. Int. J. Syst. Sci. **47**(11), 2652–2670 (2016)
258. Shayanfar, H., Gharehchopogh, F.S.: Farmland fertility: a new metaheuristic algorithm for solving continuous optimization problems. Appl. Soft Comput. **71**, 728–746 (2018)
259. Shayeghi, H., Dadashpour, J.: Anarchic society optimization based PID control of an automatic voltage regulator (AVR) system. Electr. Electron. Eng. **2**(4), 199–207 (2012)
260. Shen, H., Niu, B., Zhu, Y., Chen, H.: Optimization algorithm based on biology life cycle theory. In: International Conference on Intelligent Computing, pp. 561–570. Springer, Berlin (2013)
261. Shi, Y.: Brain storm optimization algorithm. In: International Conference in Swarm Intelligence, pp. 303–309. Springer, Berlin (2011)

262. Simon, D.: Biogeography-based optimization. IEEE Trans. Evol. Comput. **12**(6), 702–713 (2008)
263. Singh, M.K.: A new optimization method based on adaptive social behavior: ASBO. In: Proceedings of International Conference on Advances in Computing. Springer, New Delhi (2013)
264. Singh, U., Salgotra, R., Rattan, M.: A novel binary spider monkey optimization algorithm for thinning of concentric circular antenna arrays. IETE J. Res. **62**(6), 736–744 (2016)
265. Sörensen, K., Sevaux, M., Glover, F.: A history of metaheuristics. In: Handbook of Heuristics, pp. 1–18. Springer, New York (2018)
266. Storn, R., Price, K.: Differential evolution–a simple and efficient heuristic for global optimization over continuous spaces. J. Glob. Optim. **11**(4), 341–359 (1997)
267. Su, M.-C., Su, S.-Y., Zhao, Y.-X.: A swarm-inspired projection algorithm. Pattern Recognit. **42**(11), 2764–2786 (2009)
268. Subramanian, C., Sekar, A.S.S., Subramanian, K.: A new engineering optimization method: African Wild Dog Algorithm. Int. J. Soft Comput. **8**, 163–170 (2013). https://doi.org/10.3923/ijscomp.2013.163.170
269. Sun, G., Zhao, R., Lan, Y.: Joint operations algorithm for large-scale global optimization. Appl. Soft Comput. **38**, 1025–1039 (2016)
270. Sur, C., Shukla, A.: New bio-inspired meta-heuristics-green herons optimization algorithm-for optimization of travelling salesman problem and road network. In: International Conference on Swarm, Evolutionary, and Memetic Computing, pp. 168–179. Springer, Berlin (2013)
271. Sur, C., Sharma, S., Shukla, A.: Egyptian vulture optimization algorithm–a new nature inspired meta-heuristics for knapsack problem. In: The 9th International Conference on Computing and Information Technology (IC2IT2013), pp. 227–237. Springer, Berlin (2013)
272. Tabari, A., Ahmad, A.: A new optimization method: electro-search algorithm. Comput. Chem. Eng. **103**, 1–11 (2017)
273. Tahani, M., Babayan, N.: Flow Regime Algorithm (FRA): a physics-based meta-heuristics algorithm. Knowl. Inf. Syst. **60**(2), 1001–1038 (2019)
274. Taherdangkoo, M., Shirzadi, M.H., Yazdi, M., Bagheri, M.H.: A robust clustering method based on blind, naked mole-rats (BNMR) algorithm. Swarm Evol. Comput. **10**, 1–11 (2013)
275. Taillard E.D., Voss, S.: POPMUSIC—Partial optimization metaheuristic under special intensification conditions. Essays and Surveys in Metaheuristics, pp. 613–629. Springer, Berlin (2002)
276. Tamura, K., Yasuda, K.: Primary study of spiral dynamics inspired optimization. IEEJ Trans. Electr. Electron. Eng. **6**(S1), S98–S100 (2011)
277. Tan, Y., Zhu, Y.: Fireworks algorithm for optimization. In: International Conference in Swarm Intelligence, pp. 355–364. Springer, Berlin (2010)
278. Tan, K.C., Goh, C.K., Mamun, A.A., Ei, E.Z.: An evolutionary artificial immune system for multi-objective optimization. Eur. J. Oper. Res. **187**(2), 371–392 (2008)
279. Tang, D., Dong, S., Jiang, Y., Li, H., Huang, Y.: ITGO: invasive tumor growth optimization algorithm. Appl. Soft Comput. **36**, 670–698 (2015)
280. The MathWorks.: Global Optimization Toolbox, User's guide. R2018a (2018)
281. Thierens, D., Goldberg, D.: Convergence models of genetic algorithm selection schemes. In: International Conference on Parallel Problem Solving from Nature, pp. 119–129. Springer, Berlin, Oct 1994
282. Tilahun, S.L., Ong, H.C.: Prey-predator algorithm: a new metaheuristic algorithm for optimization problems. Int. J. Inf. Technol. Decis. Mak. **14**(06), 1331–1352 (2015)
283. Tkach, I., Edan, Y., Jevtic, A., Nof, S.Y.: Automatic multi-sensor task allocation using modified distributed bees algorithm. In: 2013 IEEE International Conference on Systems, Man, and Cybernetics, Manchester, pp. 1401–1406 (2013). https://doi.org/10.1109/SMC.2013.242
284. Toscano, R., Lyonnet, P.: Heuristic Kalman algorithm for solving optimization problems. IEEE Trans. Syst. Man Cybern. B Cybern. **39**(5), 1231–1244 (2009)

285. Tzanetos, A., Dounias, G.: A new metaheuristic method for optimization: sonar inspired optimization. In: International Conference on Engineering Applications of Neural Networks, pp. 417–428. Springer, Cham, Aug 2017

286. Uymaz, S.A., Tezel, G., Yel, E.: Artificial algae algorithm (AAA) for nonlinear global optimization. Appl. Soft Comput. **31**, 153–171 (2015)

287. Van den Bergh, F., Engelbrecht, A.P.: A cooperative approach to particle swarm optimization. IEEE Trans. Evol. Comput. **8**(3), 225–239 (2004)

288. Viveros Jiménez, F., Mezura Montes, E., Gelbukh, A.: Adaptive evolution: an efficient heuristic for global optimization. In: Proceedings of the 11th Annual conference on Genetic and Evolutionary Computation, pp. 1827–1828. ACM, New York, July 2009

289. Viveros-Jiménez, F., León-Borges, J.A., Cruz-Cortés, N.: An adaptive single-point algorithm for global numerical optimization. Expert Syst. Appl. **41**(3), 877–885 (2014)

290. Wang, G.G.: Moth search algorithm: a bio-inspired metaheuristic algorithm for global optimization problems. Memet. Comput. **10**(2), 151–164 (2018)

291. Wang, B., Jin, X., Cheng, B.: Lion pride optimizer: an optimization algorithm inspired by lion pride behavior. Sci. China Inf. Sci. **55**(10), 2369–2389 (2012)

292. Wang, P., Zhu, Z., Huang, S.: Seven-spot ladybird optimization: a novel and efficient metaheuristic algorithm for numerical optimization. Scientific World J. (2013). https://doi.org/10.1155/2013/378515

293. Wang, G.G., Deb, S., Gao, X.Z., Coelho, L.D.S.: A new metaheuristic optimisation algorithm motivated by elephant herding behaviour. Int. J. Bio-Inspired Comput. **8**(6), 394–409 (2016)

294. Wang, G.G., Deb, S., dos Santos Coelho, L.: Earthworm optimisation algorithm: a bio-inspired metaheuristic algorithm for global optimisation problems. Int. J. Bio-Inspired Comput. **12**(1), 1–22 (2018)

295. Wang, H., Hu, Z., Sun, Y., Su, Q., Xia, X.: Modified backtracking search optimization algorithm inspired by simulated annealing for constrained engineering optimization problems. Comput. Intell. Neurosci. (2018). https://doi.org/10.1155/2018/9167414

296. Wang, G.-G., Deb, S., Cui, Z.: Monarch butterfly optimization. Neural Comput. Appl. **31**(7), 1995–2014 (2019)

297. Wedyan, A., Whalley, J., Narayanan, A.: Hydrological cycle algorithm for continuous optimization problems. J. Optim. **2017** (2017). https://doi.org/10.1155/2017/3828420

298. Weede, O., Kettler, A., Wörn, H.: Seed throwing optimization: a probabilistic technique for multimodal function optimization. In: 2009 Computation World: Future Computing, Service Computation, Cognitive, Adaptive, Content, Patterns, pp. 515–519. IEEE, Piscataway, Nov 2009

299. Wei, Z.: A raindrop algorithm for searching the global optimal solution in non-linear programming (2013). Preprint, arXiv:1306.2043

300. Wei, Z., Huang, C., Wang, X., Han, T., Li, Y.: Nuclear reaction optimization: a novel and powerful physics-based algorithm for global optimization. IEEE Access. **7**, 66084–66109 (2019)

301. Wu, G.: Across neighborhood search for numerical optimization. Inf. Sci. **329**, 597–618 (2016)

302. Wu, H.S., Zhang, F.M.: Wolf pack algorithm for unconstrained global optimization. Math. Probl. Eng. **2014** (2014). https://doi.org/10.1155/2014/465082

303. Wu, Z., Zhao, Z., Jiang, S., Zhang, X.: PFSA: a novel fish swarm algorithm. In: Internet of Things, pp. 359–365. Springer, Berlin (2012)

304. Wu, Z., Chow, T.W., Cheng, S., Shi, Y.: Contour gradient optimization. Int. J. Swarm Intell. Res. **4**(2), 1–28 (2013)

305. Xavier, A.E., Xavier, V.L.: Flying elephants: a general method for solving non-differentiable problems. J. Heuristics **22**(4), 649–664 (2016)

306. Xi, M., Sun, J., Xu, W.: An improved quantum-behaved particle swarm optimization algorithm with weighted mean best position. Appl. Math. Comput. **205**(2), 751–759 (2008)

307. Xie, X.F., Wang, Z.J.: Cooperative group optimization with ants (CGO-AS): Leverage optimization with mixed individual and social learning. Appl. Soft Comput. **50**, 223–234 (2017)
308. Xu, Y., Cui, Z., Zeng, J.: Social emotional optimization algorithm for nonlinear constrained optimization problems. In: International Conference on Swarm, Evolutionary, and Memetic Computing, pp. 583–590. Springer, Berlin (2010)
309. Yang, X.S.: A new metaheuristic bat-inspired algorithm. In: Nature Inspired Cooperative Strategies for Optimization, pp. 65–74. Springer, Berlin (2010)
310. Yang, X.S.: Nature-Inspired Metaheuristic Algorithms. Luniver Press, Frome (2010)
311. Yang, X.S.: Flower pollination algorithm for global optimization. In: International Conference on Unconventional Computing and Natural Computation, pp. 240–249. Springer, Berlin (2012)
312. Yang, X.S., Deb, S.: Eagle Strategy Using Lévy Walk and Firefly Algorithms for Stochastic Optimization. In: González, J.R., Pelta, D.A., Cruz, C., Terrazas, G., Krasnogor, N. (eds) Nature Inspired Cooperative Strategies for Optimization (NICSO 2010). Studies in Computational Intelligence, vol 284. Springer, Berlin (2010)
313. Yang, F.C., Wang, Y.P.: Water flow-like algorithm for object grouping problems. J. Chin. Inst. Ind. Eng. **24**(6), 475–488 (2007)
314. Yang, B., Chen, Y., Zhao, Z.: A hybrid evolutionary algorithm by combination of PSO and GA for unconstrained and constrained optimization problems. In: 2007 IEEE International Conference on Control and Automation. IEEE, Piscataway (2007)
315. Yapici, H., Cetinkaya, N.: A new meta-heuristic optimizer: pathfinder algorithm. Appl. Soft Comput. **78**, 545–568 (2019)
316. Yazdani, M., Jolai, F.: Lion optimization algorithm (LOA): a nature-inspired metaheuristic algorithm. J. Comput. Des. Eng. **3**(1), 24–36 (2016)
317. Yin, P.Y., Glover, F., Laguna, M., Zhu, J.X.: Cyber swarm algorithms–improving particle swarm optimization using adaptive memory strategies. Eur. J. Oper. Res. **201**(2), 377–389 (2010)
318. Yonqkong, Z., Weirong, C., Chaohua, D., Weibo, W.: Stochastic focusing search: a novel optimization algorithm for real-parameter optimization. J. Syst. Eng. Electron. **20**(4), 869–876 (2009)
319. Yuce, B., Packianather, M., Mastrocinque, E., Pham, D., Lambiase, A.: Honey bees inspired optimization method: the bees algorithm. Insects **4**(4), 646–662 (2013)
320. Zhang, J., Sanderson, A.C.: JADE: adaptive differential evolution with optional external archive. IEEE Trans. Evol. Comput. **13**(5), 945–958 (2009)
321. Zhang, L.M., Dahlmann, C., Zhang, Y.: Human-inspired algorithms for continuous function optimization. In: 2009 IEEE International Conference on Intelligent Computing and Intelligent Systems, vol. 1, pp. 318–321. IEEE, Piscataway, Nov 2009
322. Zhang, X. et al.: Solving 0–1 knapsack problems based on amoeboid organism algorithm. Appl. Math. Comput. **219**(19), 9959–9970 (2013)
323. Zhang, X., Sun, B., Mei, T., Wang, R.: A novel evolutionary algorithm inspired by beans dispersal. Int. J. Comput. Intell. Syst. **6**(1), 79–86 (2013)
324. Zhang, H., Zhu, Y., Chen, H.: Root growth model: a novel approach to numerical function optimization and simulation of plant root system. Soft Comput. **18**(3), 521–537 (2014)
325. Zhang, Q., Wang, R., Yang, J., Lewis, A., Chiclana, F., Yang, S.: Biology migration algorithm: a new nature-inspired heuristic methodology for global optimization. Soft Comput. **23**(16), 7333–7358 (2019)
326. Zhao, S.Z., Suganthan, P.N.: Two-lbests based multi-objective particle swarm optimizer. Eng. Optim. **43**(1), 1–7 (2011)
327. Zhao, R.Q., Tang, W.S.: Monkey algorithm for global numerical optimization. J. Uncertain Syst. **2**(3), 165–176 (2008)

328. Zheng, Y.J.: Water wave optimization: a new nature-inspired metaheuristic. Comput. Oper. Res. **55**, 1–11 (2015)
329. Zheng, Y.J., Ling, H.F., Xue, J.Y.: Ecogeography-based optimization: enhancing biogeography-based optimization with ecogeographic barriers and differentiations. Comput. Oper. Res. **50**, 115–127 (2014)
330. Zielinski, K., Peters, D., Laur, R.: Stopping criteria for single-objective optimization (2005)

Chapter 3
Case Studies

3.1 Inverted Cart-Pendulum

The degree of actuation of mechanical systems must be considered when designing advanced control strategies. Indeed, mechanical systems fall into one of these three categories: *fully actuated*—the number of actuators equals the number of degrees of freedom, *over actuated*—the number of actuators is higher than the number of degrees of freedom, and *underactuated*—the number of actuators is fewer than the number of degrees of freedom [1]. Many control strategies have been designed for fully and over actuated systems [2–4]. However, these strategies are not suitable for underactuated systems.

The control of underactuated systems has inherent challenges to deal with, such as nonlinearities. However, advantages motivate the use and study of underactuated systems. For instance, since they employ fewer controllers than fully or over actuated systems, system costs and weight are lower. In case of a breakdown of actuator(s) of a fully or over actuated system, the system behaves as an underactuated one. Thus, research on developing control strategies for underactuated systems is essential.

The inverted cart-pendulum (ICP) is among the most studied underactuated systems [5]. This system represents many practical applications such as electric unicycles, two-wheeled self-balancing transporters, and a booster blastoff's orientation control. Consequently, many nonlinear control strategies deriving from the ICP are suitable for other systems in various fields in engineering, such as robotics, aerospace, and aeronautics. As a result, the ICP is a canonical benchmark problem of interest to the control community, motivating its study in this monograph.

Two steps divide its control. First, a cart is controlled to move back and forth on a rail to upswing a pendulum. Thereafter, the pendulum is maintained in its upright position. Most strategies use one controller for the swing-up action and one for the stabilization. A switchover control parameter activates either the swing-up or

© Springer Nature Switzerland AG 2021
M. J. Blondin, *Controller Tuning Optimization Methods for Multi-Constraints and Nonlinear Systems*, SpringerBriefs in Optimization,
https://doi.org/10.1007/978-3-030-64541-0_3

the stabilization controller. Many control laws have been designed to control both steps. From a survey of these laws in [6], the most cited approach to upswing the pendulum is the energy-based control method [7], and the linear-quadratic regulator (LQR) technique [8] is the most studied technique to stabilize the pendulum.

Both control laws possess parameters to tune, which influence control performance. For instance, the LQR technique employs two matrices, Q and R, which weights dictate the LQR performance. A common optimization practice fixes the R weight(s). After that, the diagonal of Q is tuned, and the parameters outside the diagonal are set to 0. Optimizing the diagonal of Q simplifies the tuning at the expense of high stabilization performance [6].

Table 3.1 surveys the publications that use the energy-based control or/and the LQR technique for the ICP control. The tuning method and the switchover control parameter are also presented. TAE stands for the trial-and-error method. Most publications focus exclusively on stabilization control. Indeed, few works on this topic consider upswing and stabilization control together, and a fewer number employs the energy-based swing-up control.

This literature survey reveals that the TAE is still a widespread technique for parameter tuning even though it is demonstrated that metaheuristics improve/facilitate the ICP controller tuning. The publications [6] and [9] are the only works that have proposed an optimization approach based on metaheuristics that optimizes at once the swing-up and stabilization controllers and the switchover control parameter. The simultaneous optimization is referred to as a holistic optimization approach. Specifically, in [6, 9], the holistic approach employs a new version of Ant Colony Optimization combined with a modified Nelder–Mead method [27]. The holistic approach considers the pendulum velocity when optimizing the switchover parameter, enabling the pendulum to reach the upright faster compared to the works that establish the pendulum angle as the switchover parameter before optimization [8, 11, 13]. Besides, the full Q matrix can be optimized in the holistic approach, enhancing stabilization performance as more degrees of freedom are kept.

Therefore, this section considers GA and SA algorithms in the holistic approach. The performances of SA and GA are studied, and the main features of metaheuristics to control tuning are demonstrated. However, the main goal of this section is twofold: (i) highlighting several elements that may influence algorithm performances and (iii) demonstrating how to deal with stability and robustness when using metaheuristic for controller optimization.

First, the influence of algorithm parametrization on algorithm performances is analyzed. To do so, different parametrizations and stop criteria are studied, and the results are compared. Along the same line, the impact of cost functions in controller tuning is studied. Besides, since SA is a trajectory-based algorithm, it will be demonstrated how initial conditions influence the algorithm performance. Also, as better solutions are sought, another essential component is to gauge the computational cost of such improvement, which is also analyzed in this section.

Second, systems tend to have parameter variations, frictions, and disturbances in a practical context. Therefore, robustness properties are considered during the optimization, which differs from all surveyed publications presented in Table 3.1. One

Table 3.1 Survey of swing-up and stabilization control tuning techniques for the ICP

Ref.	Swing-up Control	Optimizer	Stabilization LQR	Optimizer	Switchover control parameter
[6]	Energy-based	ACO-NM	Full Q	ACO-NM	Optimized angle
[9]	Energy-based	ACO-NM	Q diagonal	ACO-NM	Optimized angle
[10]	Potential-well	TAE	Q diagonal	TAE	Energy-based
[8]	PV controller	TAE	Q diagonal	TAE	Fixed angle
[11]	Kinetic energy-based	TAE	Q diagonal	TAE	Fixed angle
[12]	PD controller	TAE	Q diagonal	TAE	Fixed angle
[13]	Energy-based	TAE	Q diagonal	TAE	Fixed angle
[14]	–	–	2 terms on Q diagonal	PSO	–
[15]	–	–	Q diagonal	PSO	–
[16]	–	–	Q and **R** diagonals	Reinforced quantum-behaved PSO	–
[17]	–	–	Q diagonal	GA	–
[18]	–	–	Q diagonal	Big bang–big crunch	–
[19]	–	–	Full Q + 2PIDs	ACO	–
[20]	–	–	Full Q	Weighted artificial fish swarm algorithm	–
[21]	–	–	Full Q	ABC	–
[22]	–	–	Q diagonal	GA	–
[23]	–	–	Q diagonal	TAE	–
[24]	–	–	Q diagonal	TAE	–
[25]	–	–	Q diagonal	TAE	–
[26]	–	–	Q diagonal + PIDs	TAE	–

widespread approach is analyzing system robustness with the optimized solution, i.e., the system robustness is tested after the algorithm has provided its solutions. However, this practice is deficient. To ensure that the optimization algorithm provides a robust solution, it is mandatory to consider robustness properties during the optimization process. As a novelty, a robust optimization framework considering robustness properties based on [28] is proposed to improve the ICP control. Lastly, the ability of metaheuristic to deal with system restrictions will be demonstrated.

Section 3.1.1 presents the ICP model and its controller structures. The controller structure optimization is described in Sect. 3.1.2. Simulation results and analyses are given in Sect. 3.1.3 followed by a discussion in Sect. 3.1.4.

3.1.1 Model of the Inverted Cart-Pendulum and its Controllers

The ICP modeled in this monograph is the IP02 Linear Inverted Pendulum from Quanser Inc. [29]. Table 3.2 presents the variables and constants of the system [29]. Figure 3.1 presents the diagram of the ICP system with its control system.

Table 3.2 Parameters and variables of the ICP system

Variable	Definition	Unit
x_c	Cart position	m
$\dot{x}_c$	Cart velocity	m/s
$\ddot{x}_c$	Cart acceleration	m/s^2
$x_{c_{mea}}$	Measured cart position	m
$\dot{x}_{c_{mea}}$	Measured cart velocity	m/s
F_c	Linear force applied to the cart generated by the servo motor	N
V_m	Servo voltage, control input	V
α	Pendulum angle	rad
$\dot{\alpha}$	Pendulum angle velocity	rad/s
$\ddot{\alpha}$	Pendulum angle acceleration	rad/s^2
α_{mea}	Measured pendulum angle	rad
$\dot{\alpha}_{mea}$	Measured pendulum angle velocity	rad/s
g	Gravity constant	9.81 m/s^2
l_p	Distance from the pivot to the pendulum gravity center	0.3302 m
r_{mp}	Motor pinion radius	6.35×10^{-3} m
$x_{c_{lim}}$	Cart limit position	± 0.35 m
$\dot{x}_{c_{lim}}$	Cart limit velocity	± 4 m/s
B_{eq}	Equivalent viscous damping coefficient as seen at the motor pinion with M_w	0.94 N s/m
B_p	Viscous damping coefficient as seen at the pendulum axis	0.0024 N s/rad
J_{eq}	Lumped mass of the cart system with M_w	1.0431 kg
J_p	Pendulum moment of inertia at its center	0.0079 kg m^2
K_g	Planetary gearbox gear ratio	3.71
K_m	Motor back-emf constant	7.68×10^{-3} V s/rad
K_t	Motor current–torque constant	7.68×10^{-3} N m/A
L_p	Total length of the pendulum	0.6413 m
L_t	Track length	0.8 m
M_c	Mass of cart with 3 cables connected plus mass for self-erecting experiment	0.94 kg
M_p	Pendulum mass	0.2300 kg
R_m	Motor armature resistance	2.6 Ω
u	ICP control signal	V
u_{swing}	Upswing control signal	V
u_{stab}	Stabilization control signal	V
$V_{m_{max}}$	Servo voltage saturation	± 5 V
η_g	Planetary gearbox efficiency	1.0
η_m	Motor efficiency	0.95

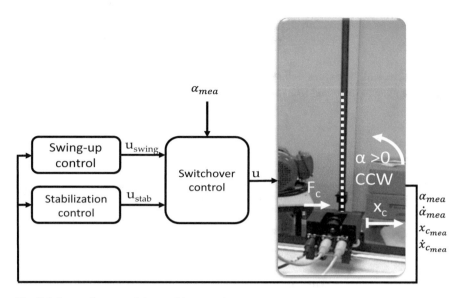

Fig. 3.1 Inverted cart-pendulum and its control system

A pendulum is joined to a cart with a pole shaft. No actuator moves the pendulum. Indeed, it is the cart displacement powered by a DC motor that makes the pendulum oscillate. The switchover control decides based on α_{mea} whether u_{swing} or u_{stab} controls the cart. α takes a positive value when the pendulum rotates in a counterclockwise direction (CCW). When the cart travels from its initial position to the right, x_c is positive. Restrictions are added to the cart position. In particular, the cart must return to its initial position for the stabilization control, and the cart displacement must respect the track length L_t.

In robotics, the Lagrange model is often used to describe more complex systems. Therefore, the following nonlinear equations of motion model the dynamics of the ICP: (Please see [30] for more information on the origin of the equations.)

$$(J_{eq} + M_p)\ddot{x}_c + M_p l_p \cos(\alpha)\ddot{\alpha} - M_p l_p \sin(\alpha)\dot{\alpha}^2 = F_c - B_{eq}\dot{x}_c \tag{3.1}$$

$$M_p l_p \cos(\alpha)\ddot{x}_c + (J_p + M_p l_p^2)\ddot{\alpha} + M_p l_p g \sin(\alpha) = -B_p\dot{\alpha} \tag{3.2}$$

The F_c, J_p, and J_{eq} are obtained with the following equations:

$$F_c = \left(\frac{\eta_g K_g K_t}{R_m r_{mp}}\right)\left(\frac{-K_g K_m \dot{x}_c}{r_{mp}} + \eta_m V_m\right) \tag{3.3}$$

$$J_p = \frac{M_p L_p^2}{12} \tag{3.4}$$

$$J_{eq} = M_c + \frac{\eta_g K_g^2 J_m}{r_{mp}^2} \tag{3.5}$$

With the state variables defined as $x_1 = x_c$, $x_2 = \alpha$, $x_3 = \dot{x}_c$, and $x_4 = \dot{\alpha}$, the resulting system (3.1) and (3.2) is

$$\dot{x}_1 = x_3 \tag{3.6}$$

$$\dot{x}_2 = x_4 \tag{3.7}$$

$$\dot{x}_3 = \frac{(F_c - B_{eq} x_3 + M_{pl_p} \sin(x_2) x_4^2)(J_p + M_{pl_p^2}) - (B_p x_4 - M_{pl_p} g \sin(x_2))(M_{pl_p} \cos(x_2))}{(J_{eq} + M_p)(J_p + M_{pl_p^2}) - (M_{pl_p} \cos(x_2))^2} \tag{3.8}$$

$$\dot{x}_4 = \frac{(J_{eq} + M_p)(-B_p x_4 - M_{pl_p} g \sin(x_2)) - (M_{pl_p} \cos(x_2))(F_c - B_{eq} x_3 + M_{pl_p} \sin(x_2)(x_4)^2)}{(J_{eq} + M_p)(J_p + M_{pl_p^2}) - (M_{pl_p} \cos(x_2))^2} \tag{3.9}$$

3.1.1.1 Controller Structures

Swing-Up Control
The upswing control law is derived from the pendulum energy. By ignoring friction and taking the cart acceleration as a control input, the following equation is obtained from (3.2) [7]:

$$(J_p + M_{pl_p^2})\ddot{\alpha} + M_p g l_p \sin(\alpha) = -M_{pl_p} u \cos(\alpha) \tag{3.10}$$

where u is the cart acceleration. F_c in relation with u is estimated by

$$F_c = J_{eq} u \tag{3.11}$$

where the required control voltage is calculated with (3.3). The pendulum potential energy E_p is defined by

$$E_p = M_p g l_p (1 - \cos(\alpha)) \tag{3.12}$$

The pendulum kinetic energy E_k is computed with

$$E_k = \frac{1}{2}(J_p + M_{pl_p^2})\dot{\alpha}^2 \tag{3.13}$$

and the pendulum total energy is $E_t = E_p + E_k$.

From (3.12) and (3.13), and the derivative of E_t, the following energy-based control law is obtained:

$$u = sat_{umax}(\mu(E_t - E_{ref})\text{sign}(\dot{\alpha}\cos(\alpha))) \tag{3.14}$$

where μ is a tunable parameter, sat_{umax} is the control saturation coefficient, and E_{ref} refers to the highest E_p value. The stability proof of (3.14) is available in [7]. The upswing control starts when $\dot{\alpha} \neq 0$, because if $\dot{\alpha} = 0$, $u = 0$.

Stabilization Control

The stabilization controller design with the LQR technique requires a linearized model around the vertical pendulum position. The linearized model is defined as follows:

$$\dot{x}_p = Ax_p + BV_m \tag{3.15}$$

$x_p = [x_c, \alpha - \pi, \dot{x}_c, \dot{\alpha}]$ corresponds to the pendulum upper vertical variations around its equilibrium point, $x = [0, \pi, 0, 0]$.

Computed from (3.3) and (3.6)–(3.9), A and B at $x = [0, \pi, 0, 0]$:

$$A = 1/J_T \begin{bmatrix} 0 & 0 & 1 & 0 \\ 0 & 0 & 0 & 1 \\ 0 & M_p^2 l_p^2 g & -B_{eq}(J_p + M_p l_p^2) & -M_p l_p B_p \\ 0 & (J_{eq} + M_p)M_p l_p g & -B_{eq}M_p l_p & -(J_{eq} + M_p)B_p \end{bmatrix} \tag{3.16}$$

$$B = 1/J_T \begin{bmatrix} 0 \\ 0 \\ (J_p + M_p l_p^2) \\ (M_p l_p) \end{bmatrix} \tag{3.17}$$

with

$$J_T = (J_{eq} + M_p)J_p + J_{eq}M_p l_P^2 \tag{3.18}$$

Given the actuator dynamics, (3.16) and (3.17) become

$$A(3, 3) = A(3, 3) - B(3)\eta_g K_g^2 \eta_m KtKm/r_{mp}^2/R_m$$

$$A(4, 3) = A(4, 3) - B(4)\eta_g K_g^2 \eta_m KtKm/r_{mp}^2/R_m \tag{3.19}$$

$$B = \eta_g K_g \eta_m K_t/r_{mp}/R_m B$$

The LQR technique optimizes the following function:

$$J_{LQR} = \int_0^\infty (x_p^T Qx_p + V_m^T RV_m)dt \tag{3.20}$$

and it computes the optimal gain vector K of the state-feedback control law:

$$V_m = -Kx_p \tag{3.21}$$

which is subject to the linearized state dynamic (3.15).
Q and R are defined as

$$Q = \begin{bmatrix} q_1 & q_5 & q_6 & q_8 \\ q_5 & q_2 & q_7 & q_9 \\ q_6 & q_7 & q_3 & q_{10} \\ q_8 & q_9 & q_{10} & q_4 \end{bmatrix}, R = \begin{bmatrix} r_1 \end{bmatrix} \tag{3.22}$$

where q_1 to q_{10} and r_1 are tunable gains.

3.1.2 Controller Structure Optimization

The holistic optimization tunes simultaneously the ICP controller parameters
through the following optimization problem [6]:

$$\underset{x}{\text{minimize}} \ f(x) = \int_t^{T_{switch}} t \, dt + \sigma_1 \int_{T_{switch}}^{t_s} (t - T_{switch})(|\alpha - \alpha_{ref}| + |\dot{x}_c|) dt$$
$$+ \sigma_2 \int_0^{t_s} \Delta_{x_c} dt \tag{3.23}$$

where

$$\Delta_{x_c} = \begin{cases} |x_c| - x_{c_{lim}} \ \textbf{\textit{if}} \ |x_c| > x_{C_{lim}} \\ 0 \qquad\qquad otherwise \end{cases} \tag{3.24}$$

where x is the decision vector corresponding to the controller parameters. $x_{C_{lim}}$
is a soft constraint on the cart position, σ_1 and σ_2 are weights, α_{ref} refers to the
pendulum vertical position (pointing upward) ($\pm\pi$ rad), t_s is the simulation time,
and T_{switch} is the time when the stabilization controller takes over the swing-up
controller. The first term in (3.23) is the integral of time from the beginning of the
simulation until the stabilization controller takes over. This term is to reduce the
swing-up time. For high-level stabilization performance, pendulum movements and
cart velocity are penalized through the second term in (3.23). Also, the cart velocity
penalization is to fulfill the cart position requirement during stabilization control,
i.e., the cart has to be at its initial position when stabilization control takes over.
The last term is the soft constraint on the cart position, in which the optimization
penalizes the solutions for cart displacements beyond $x_{C_{lim}}$. Considering the track
length ensures that the algorithm reaches an optimized solution that is viable in
practice. In short, the cost function considers the cart-track length, minimizes the
swing-up time, and assures high stabilization performance. For the ICP understudy,
$\sigma_1 = \sigma_2 = 1,000$.

Q must be a symmetric positive semidefinite matrix to ensure that the LQR technique works. The Q is defined as symmetric in the problem statement. Therefore,
evaluating that its eigenvalues are nonnegative assures the positive semidefiniteness

of Q [31]. If Q is not positive semidefinite or if the optimized parameter set results in unstable system response, the solution is penalized during the optimization process, i.e., a high-cost function value is assigned to this solution.

For the ICP understudy, $x = [\mu \ sat_{umax} \ \epsilon \ q_1 \ q_2 \ q_3 \ q_4 \ q_5 \ q_6 \ q_7 \ q_8 \ q_9 \ q_{10}]$ and the lower and upper bounds are for the search:

$$\mu = [350 \text{ to } 550] \quad k_{cw} = [0.5 \text{ to } 10] \ k_{vm} = [0.5 \text{ to } 10]$$
$$k_{em} = [0.5 \text{ to } 15] \quad \eta_{mw} = [0.8 \text{ to } 2] \ k_{su} = [0.5 \text{ to } 10]$$
$$sat_{umax} = [2 \text{ to } 10] \ \epsilon = [0.05 \text{ to } 0.4] \ q_1 = [10 \text{ to } 1,000]$$
$$q_2 = [10 \text{ to } 1,000] \ q_3 = [0 \text{ to } 35] \quad q_4 = [0 \text{ to } 35]$$
$$q_5 = [-15 \text{ to } 15] \quad q_6 = [-15 \text{ to } 15] \ q_7 = [-15 \text{ to } 15]$$
$$q_8 = [-15 \text{ to } 15] \quad q_9 = [-15 \text{ to } 15] \ q_{10} = [-15 \text{ to } 15]$$

The switchover control of most ICP published research occurs when the pendulum is near its upright position with low velocity, usually $\pm 5°$. On the contrary, the ϵ optimization with the holistic approach considers the pendulum velocity, which allows the pendulum to reach its upright earlier with a faster velocity.

3.1.3 Simulation Results

The GA algorithm is available in [32], and the SA algorithm is available in the MATLAB optimization toolbox. The nonlinear mathematical model of the ICP is designed on Simulink^{TM}. The ICP model neglects the dry friction. From x_c and α, $\dot{x}_c$ and $\dot{\alpha}$ are, respectively, calculated with a high-pass filter as follows:

$$\frac{w_{cf}^2}{s^2 + 2\zeta w_{cf} + w_{cf}^2} \tag{3.25}$$

where w_{cf} is the filter cutting frequency and ζ the filter damping ratio. A 1 V signal initiates the cart movement. When $\dot{\alpha} > 0.1$ rad/s, the control sequence starts.

Influence of the Stop Criterion on Algorithm Performances

This set of simulation scenarios highlights the impact of the stop criterion on algorithm performances. Table 3.3 presents the simulation results, where GA and SA paramets are set to their default values; GA—number of individual, NIND=50, crossover probability, Pc=0.9 and mutation probability, Pm=0.1 [33] and SA—reanneal interval, ReanInt = 100 and initial temperature, InitTemp = 100. Setting algorithm parameters to their default values is a suitable and meaningful choice given that algorithm fine-tuning might be a costly process and might not provide significantly better results [34]. The GA initial population and the SA starting parameters are created randomly. The algorithm stop criterion st varies by taking 2000, 5000, 10,000, and 100,000 as a maximum number of function evaluations.

Table 3.3 Simulation results obtained by GA and SA: default parametrization for different stop criterion values

Scenario	I-A		I-B		I-C		I-D	
st	2000		5000		10,000		100,000[a]	
Algorithm	SA	GA	SA	GA	SA	GA	SA	GA
ME	158.24	84.15	128.97	77.21	132.67	76.14	89.83	73.66
SD	37.44	4.29	21.13	1.91	28.47	1.09	–	–
$f - evals$	2,000	1,940	5,000	4,850	9,674	9,720	70,878	99,600

[a]Only one run of each algorithm was performed for this stop criterion due to long simulation time

ME refers to the average of $f(x)$ values. SD is the average of the standard deviations. $f - evals$ is the average of the number of $f(x)$ evaluations to reach the best solution. The average is computed over ten runs.

A widespread practice is assessing algorithm performances based on the balance between the precision and the computational cost measured by ME and $f - evals$, respectively. Table 3.3 shows that ME value tends to decrease as the st increases. ME varies from 158.24 to 89.83 with SA and from 84.15 to 73.66 with GA. Results obtained with the SA algorithm in scenario **I-B** are better than those obtained in scenario **I-C**. These results might be surprising since a higher number of function evaluations as a stop criterion gives the algorithm more attempts to reach the optimal solution. However, those results suggest that the SA algorithm is strongly affected by its initial state and parametrization that is not finely adjusted. Indeed, suppose the decision vector is too far from the optimal solution. In that case, the algorithm might not be able to reach the area of the optimal solution with an unfine-tuned parametrization. The impact of the initial state on SA performance is studied later in this section. Nonetheless, simulation results support that the stop criterion influences more SA performance than GA performance. Consistently, the SD values are significantly high with SA and low with GA, which corroborates that SA performance strongly depends on the initial conditions and parametrization compared to GA.

For computational cost and time purposes, the optimization design is often optimized so that satisfactory results are reached as fast as possible. Speed becomes a more crucial feature in online optimization. The optimized solution might not be the best reachable by the algorithm, but the solution is sufficiently satisfying to be implemented. For instance, GA results show that ME is 2.48 lower with $st = 100,000$ compared to $st = 10,000$, which represents an improvement of 3.37 % on ME at the expense of a computation cost 10 times higher. If the ME improvement is crucial for the problem at hand, it is worth increasing the computation cost. Otherwise, it is preferable to keep the computational cost low. From the obtained results with SA and GA, the best compromise between computational cost and precision is reached at $st = 10,000$.

Comparing algorithm performances, GA outperforms SA for the four tested stop criteria. Indeed, GA has a ME value that is lower for all scenarios. In addition, GA has a lower $f - evals$ value for $st = 2,000$ and $st = 5,000$. For $st = 10,000$ and $st = 100,000$, GA obtained a higher $f - evals$ value, but ME is significantly lower, making the higher computational cost worth it. From an optimization point of view, a lower ME always means a better solution. However, it is important to verify the meaning of this improvement on the system response to ensure that the cost function leads the optimization process to desired system responses. Figures 3.2, 3.3, 3.4, and 3.5 show the best ICP responses according to the ME value for scenarios I-A to I-D, respectively.

Regardless of the algorithm, the pendulum positions are mostly the same for all scenarios. However, the cart position differs, which explains the ME differences between GA and SA. GA reaches a better ME value because the cart moves less at the beginning of the stabilization compared to SA. Since this feature was a constraint on the cart during stabilization, the difference on ME is meaningful. Nonetheless, the cart position restrictions are met for all scenarios; the cart remains within the track length limits L_t. Therefore, the control designer must evaluate the meaning of a better ME and decide if the optimization process is meaningful. In a practical context with online control tuning, not only the optimized solution matters but also the computational costs.

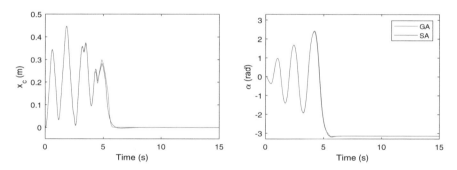

Fig. 3.2 ICP responses obtained by GA and SA for scenario I-A

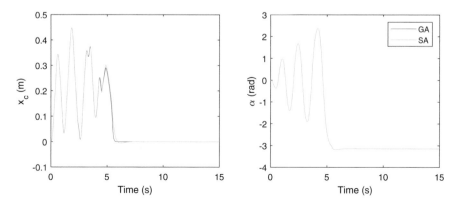

Fig. 3.3 ICP responses obtained by GA and SA for scenario I-B

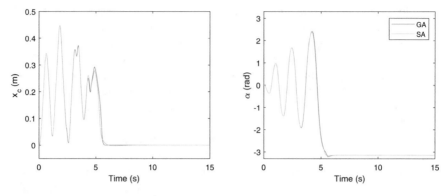

Fig. 3.4 ICP responses obtained by GA and SA for scenario I-C

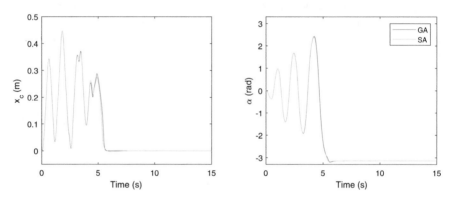

Fig. 3.5 ICP responses obtained by GA and SA for scenario I-D

Influence of Parametrization on Algorithm Performances

The second set of simulation scenarios is performed with different parametrizations for a pre-established stop criterion, i.e., $st = 10,000$. This stop criterion allows the algorithm to reach satisfactory results for a decent computational cost. Table 3.4 presents the simulation results obtained with SA and GA along with the parameter value that has been changed in the parametrization. The remaining parameters are set to their default values. Therefore, the results can only be compared to scenario **I-C** because the stop criterion is the same.

The parametrization modifications improve SA performance compared to scenario **I-C**. The best ME value is reached when ReanInt = 50 instead of 100- its default value. This result suggests that the reannealing process- used to help the algorithm getting out of potential local minima-is important for the application. Specifically, the reannealing process raises the temperature sooner and more often when ReanInt = 50. Regarding GA, the algorithm reached better results with a higher mutation rate than its default value-Pm=0.3 instead of 0.1. In particular, ME and $f - evals$ are lower when Pm=0.3 compared to GA results of scenario

Table 3.4 Simulation results obtained by GA and SA with different parametrizations for $st = 10,000$

Scenarios	SA			GA		
	II-A	II-B	II-C	II-D	II-E	II-F
Changes	InitTemp = 150	InitTemp = 50	ReanInt = 50	NIND=100	Pc=0.7	Pm=0.3
ME	125.05	127.95	120.59	76.20	77.1320	75.73
SD	26.74	29.69	25.86	1.65	1.73	1.0387
$f - evals$	8,751	10,000	9,787	9,780	9,885	9,045

I-C. The mutation action is employed to find better solutions that do not already exist within the population and cannot be created through other GA actions. This action helps GA to escape local minima. SA and GA needed more intervention to escape local minima. These two sets of simulation scenarios demonstrate the influence of parametrization on algorithm performances. Please take a note that the best parametrization for SA and GA is not assumed. However, it highlights the importance of trying different sets of parameters to improve results and draw conclusions regarding an algorithm's performance applied to optimization problems.

For ICP control tuning, the GA algorithm beats the SA algorithm for all scenarios. Also, SD value indicates that optimizing with GA has a higher probability of reaching an optimized solution near the obtained ME. Therefore, for the ICP understudy, GA is superior to SA.

Impact of the Cost Function

The performance of metaheuristics for control tuning also relies on the cost function designed for the system at hand. As mentioned in Sect. 2.3, ITAE is a typical performance criterion used as a cost function. To analyze the impact of a cost function on the optimized solution, GA, with the same parametrization as scenario **II-F**, optimizes the ICP controller with the ITAE. The obtained solution is compared to scenario **II-F**. The ITAE is computed on the pendulum and cart position as follows:

$$f_{ITAE}(x) = \int_0^{t_s} t|\alpha - \alpha_{ref}|dt + \int_0^{t_s} t\Delta_{x_c}dt \tag{3.26}$$

Figure 3.6 compares the ICP responses obtained with (3.23) and (3.26).

The pendulum reaches the upright position in one less swing with (3.26) compared to (3.23). However, the cart requires significantly more time to return to its initial position for the stabilization control. Therefore, the comparison suggests not only that the ICP response obtained with (3.23) presents a better balance between swing-up time, stabilization performance, and cart restriction fulfillment but also that designing a cost function should be taken seriously.

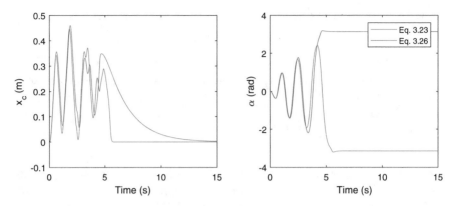

Fig. 3.6 ICP responses comparisons—different cost functions

With this in mind, it is mandatory to analyze if the user-defined cost function suits the problem to optimize before making any conclusions regarding the algorithm performance or the controller adequacy for controlling the system.

Impact of Initial Conditions on SA Performance

As mentioned earlier, the SA algorithm is a trajectory-based algorithm. The optimization starts from an initial state, i.e., a controller's parameter. Thus, it suggests that the algorithm performance depends on the initial state condition. Suppose the initial state condition is far away from the optimal solution, and the objective function is non-convex. The algorithm may not be able to converge in the region of the optimal solution. Fine-tuned algorithm may improve results. However, algorithm fine-tuning may result in solving another optimization problem [34]. Nevertheless, to expose this dependency and elaborate on high SD values obtained previously, ten simulation runs with different initial state conditions are performed. The SA parametrization is set to its default values and $st = 10,000$. Table 3.5 presents the initial state conditions (I.C.) for each run and the final results along with the objective function values. The notation $-$ is assigned to $f(x)$ when the state vector x leads to a nonviable solution. Different reasons motivate the selection of each I.C. The first I.C. is one of the optimized solutions obtained by GA in scenario **I-C**. The second I.C. is half of the first I.C. parameter set value, except for μ to respect the lower bound. The 3rd and 4th I.C. sets are the lower and upper bounds, respectively, while the 5th set is halfway between the bound values. The 6th, 7th, 8th, and 9th I.C. are values generated randomly between the lower and upper bounds. The 10th I.C. is the parameter obtained by GA in scenario **I-D**.

The optimized solutions vary greatly; $ME = 134.54$ and $SD = 39.49$. The only changing element between runs is the I.C.. This corroborates that the SA convergence to the (near-)optimal solution is extremely related to its I.C. Contrary to all other runs, the SA algorithm could not improve the solution of run 10, which is the solution obtained with GA in scenario **I-D**. This means either the solution is

Table 3.5 Different initial conditions for SA algorithm

		μ	sat_{umax}	ϵ	q_1	q_2	q_3	q_4	q_5	q_6	q_7	q_8	q_9	q_{10}	$f(x)$
1	I.C.	548.63	6.49	0.34	973.42	226.07	19.50	2.01	12.23	−9.02	−3.26	1.23	2.02	−6.25	77.44
	Results	548,66	6.61	0.13	973.42	226.02	19.67	1.98	12.15	−9.01	−3.31	1.22	2.03	−6.17	74.83
2	I.C.	300	3.24	0.17	486.71	113.03	9.75	1.01	6.11	−4.51	−1.63	0.61	1.01	−3.12	–
	Results	329.99	6.72	0.18	450.64	90.72	32.22	33.51	9.89	8.96	−7.20	13.11	8.79	−11.99	150.48
3	I.C.	300	2.00	0.05	10.00	10.00	0.00	0.00	−15.00	−15.00	−15.00	−15.00	−15.00	−15.00	–
	Results	317.38	6.74	0.20	58.36	40.75	30.48	24.54	−1.82	−6.60	0.16	14.97	−6.83	−14.99	182.28
4	I.C.	550	10.00	0.40	1,000.00	1,000.00	35.00	35.00	15.00	15.00	15.00	15.00	15.00	15.00	929.31
	Results	547.97	6.65	0.14	998.70	996.27	33.74	29.20	−1.00	7.67	13.11	7.23	11.80	10.09	124.57
5	I.C.	425	6.00	0.23	505.00	505.00	17.50	17.50	0.00	0.00	0.00	0.00	0.00	0.00	343.42
	Results	423.17	6.62	0.21	505.25	498.04	18.36	11.53	5.32	−4.23	8.30	8.85	2.32	0.69	119.40
6	I.C.	503.68	9.25	0.09	914.24	636.04	3.41	9.75	1.41	13.73	13.95	−10.27	14.12	13.72	–
	Results	474.93	9.73	0.36	917.23	573.39	7.22	16.09	3.67	0.56	−0.90	0.29	−1.60	9.49	188.07
7	I.C.	421.34	8.40	0.10	428.54	916.58	27.73	33.58	4.67	−13.93	10.47	13.02	5.36	7.73	–
	Results	425.45	6.66	0.19	428.47	909.92	30.17	18.46	−11.21	7.86	−6.15	14.35	−0.41	13.09	146.95
8	I.C.	485.78	5.14	0.28	179.47	708.99	1.11	9.69	−13.61	−12.09	9.70	5.84	−5.49	13.51	–
	Results	462.11	6.65	0.22	178.17	700.89	23.60	19.95	−1.04	−9.66	7.45	7.02	6.47	3.38	161.97
9	I.C.	535.93	8.98	0.32	372.20	629.37	33.88	33.48	3.88	13.26	6.64	6.26	−9.05	−13.94	305.57
	Results	507.23	6.65	0.15	369.55	636.01	25.78	26.36	−8.80	2.27	11.34	13.75	8.62	−14.69	123.25
10	I.C.	549.50	6.54	0.20	997.43	346.33	33.84	6.65	15.00	14.09	5.52	−6.83	−2.82	−15.00	73.66
	Results	549.50	6.54	0.20	997.43	346.33	33.84	6.65	15.00	14.09	5.52	−6.83	−2.82	−15.00	73.66

in an optimal local area, and the algorithms are trapped in this area, or the solution is the optimal solution. It is worth mentioning that the optimized switchover control parameter ϵ equals $11.46°$. This angle value is significantly higher than most works in the subject; usually, the switchover angle is around $5°$. It is also interesting to analyze that run 1 has a $f(x)$ value close to run 10, but the SA algorithm could not reach the same final results. Indeed, the $f(x)$ values are really close, but the controller values differ greatly for some parameters. It proposes that the objective function has local optima of similar $f(x)$ values, and the SA algorithm could not exit the optimum local region.

Even though SA has a specific strategy to escape from local minima [35], results suggest that combining SA with a population-based metaheuristic is a promising, if not necessary, alternative. For more details about SA-based hybrid algorithms, please see [35] and Chap. 4.

Control Tuning and Robustness Properties

Robustness properties are important in control. Paradoxically, a limited number of works on control tuning by metaheuristics consider robustness during the optimization process. Among them, [36] considers stability and robustness criteria in the cost function exclusively. However, neglecting dynamic response can lead to poor dynamic system performance. On the other hand, cost functions with time and frequency domain specifications seem more promising [37]. But it often increases the number of tunable weights within the cost function, becoming cumbersome to tune. Therefore, in this monograph, a robust optimization framework based on the recent publication [28] that possesses a low number of tunable weights is proposed. The optimization framework allows considering robustness properties during the optimization as well as dynamic performances. The proposed robust optimization framework evaluates (3.23) with the *robustness assessment* used in [9]. In particular, the cost function (3.23) is optimized with the ICP model with $\pm 25\%$ and $\pm 50\%$ uncertainties on ICP parameters prone to variations. The selected parameters are J_{eq} and B_{eq}. The cost function average $\overline{f(x)}$ now the objective function. Performing the *robustness assessment* only on the critical elements allows the optimization to reach a robust solution for a low computational cost.

As GA is better than SA, only GA is employed in the robust optimization framework. GA parametrization is the same as scenario **II-F** and $st = 10,000$. Table 3.6 presents the values of the controller parameters along with $\overline{f(x)}$. The control structure A is the parameter set obtained with the robust optimization framework. The controller structure B is the parameter set that has the lowest ME value in scenario **II-F**. Figures 3.7 and 3.8 present the ICP responses obtained with controllers A and B, respectively, as J_{eq} and B_{eq} vary.

The robust optimization framework allows GA to reach an optimized controller structure robust to B_{eq} and J_{eq} variations. In contrast, controller B cannot handle all variations. Moreover, results suggest that a high switchover control parameter value improves robustness properties. In particular, ϵ equals $17.19°$ for controller A. To draw stronger conclusions about controller A robustness properties, simultaneous

Table 3.6 Comparison of simulation results obtained with GA - controller A vs. B

Controller	μ	sat_{umax}	ϵ	q_1	q_2	q_3	q_4	q_5	q_6	q_7	q_8	q_9	q_{10}	$\overline{f(x)}$
A	535.9	9.0	0.3	372.2	629.4	33.9	33.5	3.9	13.3	6.6	6.3	−9.1	−14.0	536.4
B	541.0	6.55	0.19	1,000	343.0	34.5	6.7	14.2	1.3	3.7	−7.0	−8.8	−15	55,619

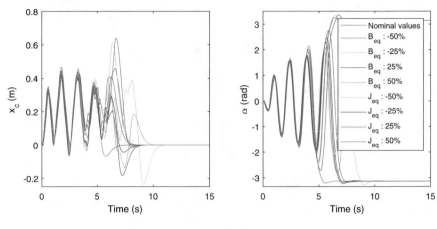

Fig. 3.7 ICP responses obtained with controller A

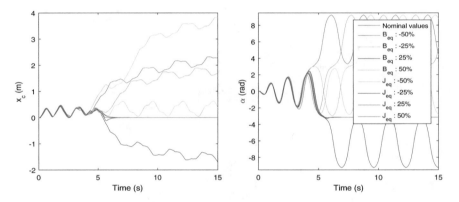

Fig. 3.8 ICP responses obtained with controller B

random variations are applied from a range of $\pm50\%$ of the nominal values, which are $J_{eq} = 1.0431$ and $B_{eq} = 0.94$. Table 3.7 presents the ICP parameter values, while Fig. 3.9 shows the ICP responses obtained with controller A.

The robust optimization framework allows the optimization algorithm to reach a parameter set robust to simultaneous random parameter system variations, proving the proposed framework efficiency.

Another critical feature in control is the system's ability to deal with disturbances. Consequently, a disturbance is added to the control signal at 10 s. Figure 3.10 shows the ICP responses obtained.

Controller A manages the disturbance, while controller B does not. According to the results obtained, the proposed robust optimization framework provides an

Table 3.7 Simultaneous ICP
parameter variations

Variations	B_{eq}	J_{eq}
A	1.218	0.714
B	1.372	1.089
C	0.600	0.675
D	0.712	1.394
E	0.709	1.367
F	0.699	1.486
G	0.799	0.725
H	0.706	1.161
I	0.915	0.886
J	1.251	1.129

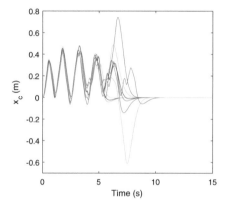
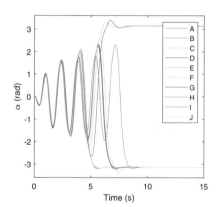

Fig. 3.9 ICP responses obtained with controller A under simultaneous random variations on B_{eq} and J_{eq}

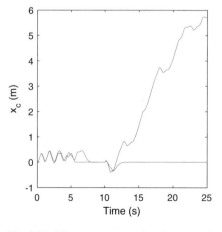
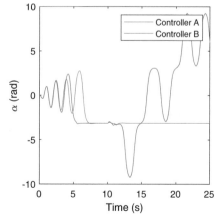

Fig. 3.10 ICP responses to a disturbance at 10 s

optimized controller parameter set that presents robustness properties; the optimized control system copes with simultaneous parameter variations and disturbances.

3.1.4 Discussion

In this section, we have seen that metaheuristics performances for control tuning depend on several "external" factors of the algorithm itself. In particular, the algorithm parametrization is an influential factor. Default parameter values are often a reasonable choice [34]. However, inappropriate parametrization drives the algorithm to unsatisfactory solutions that may lead to wrong conclusions regarding the algorithm's suitability or the controller adequacy for the system to control. In addition, the user must define an appropriate cost function for the system at hand. If an inadequate performance criterion is designed, the algorithm will not converge to desired results no matter how powerful the optimization algorithm is.

The advantages of metaheuristics for controller tuning have been demonstrated. One major feature that other classes of optimization techniques do not always have is the user-defined cost function that allows targeting system requirements during the optimization. Moreover, robustness properties can also be considered during the optimization so that the optimized solution possesses robust control properties. In particular, the first holistic approach combining a robust optimization framework for controller tuning by metaheuristic was proposed for the ICP. Simulation results substantiate the efficiency of the proposal. Indeed, the optimized controllers provide satisfactory system responses while coping with ICP parameter variations and disturbances. Moreover, it has been demonstrated that the approach offers an optimization that is controller-independent. In particular, SA and GA could optimize the energy-based controller and the LQR controller. It is noteworthy to recall that the LQR technique is an optimal control method, which is usually tuned by trial-and-error. Metaheuristics not only improve its tuning but also provide greater control performance by facilitating optimizing the full Q instead of only its diagonal.

Based on the final objective function value, the results show that GA, a population-based algorithm, offers greater performance compared to SA, a trajectory-based algorithm. Nonetheless it is important to analyze the meaning of a lower-cost function value on the system response. For instance, for scenario **I-D**, the cost function value is 97.61 for SA and 73.66 for GA, which represents an improvement of 32%. Figure 3.5 shows that this 32% corresponds to the cart reaching its initial position quicker during stabilization control. Since this characteristic was a system requirement, this improvement is significant. But if the improvement were negligible, the system's cost function suitability would have been questionable. A better cost function value should always result in a better system response according to the system requirements.

3.2   AVR System

The control of the AVR system has been deeply studied over the past two decades. The majority of works on the subject propose to control the AVR system with a PID-based controller optimized by metaheuristics. Due to a high number of publications on the subject, the AVR system became a canonical benchmark for testing new metaheuristics in control. Table 3.8 presents an exhaustive review of the proposed metaheuristics to optimize PID-based controllers for the AVR system. The survey presents the type of PID-based controller along with the proposed optimizer. The last column of Table 3.8 indicates the algorithms, works, or controllers with which the work presented in the first column is compared to.

As this survey shows, most publications compare the performance obtained with the proposed optimizer to solutions obtained with other metaheuristics. This comparison is used to validate new algorithms' efficiency. Nonetheless, metaheuristics are also a great tool to compare the performance of different controllers. Indeed, to compare controllers' performance fairly, it is advised to apply the same tuning approach. In doing so, it reduces the risk of performance bias that different tuning methods could introduce. Therefore, in this section, GA and SA optimize different controllers to compare their performances. The filtered FOPID (FOPID$_f$), PID-Acceleration (PIDA), and the filtered two-degree-of-freedom-PID (2DOFPID$_f$) controllers have been selected because a sparse number of publications or none employs these structures to control the AVR system. These controllers have the same number of tunable parameters. Thus, the algorithm parametrization can remain identical as algorithm parametrization setting relies, in part, on the number of variables to optimize. As a result, it decreases the risk of comparative bias that different algorithm parametrization could cause. For instance, the size of the population of population-based algorithms depends on the number of parameters to optimize. Typically, a higher number of parameters to optimize necessitate a bigger population. Therefore, the GA and SA parametrizations can remain constant for FOPID$_f$, 2DOFPID$_f$, and PIDA optimization without biasing the performance results.

In addition to controller comparison, this section analyzes another aspect that affects algorithm performances. Control designers should pay attention to the search space defined to solve the problem at hand since it influences the algorithm performance and, consequently, the controller performance. In controller tuning, parameter bounds must be provided for several reasons. For example, in [83], negative values are prohibited to assure system stability and robustness. However, it is sometimes challenging to determine the search space when the control designer lacks knowledge of potential parameters' values. Thus, biases in the controller performance comparison could be caused by the search space. Although the number of tunable parameters is the same for FOPID$_f$, 2DOFPID$_f$, and PIDA controllers, it does not mean that the parameter values are similar. To reduce as much as possible comparison biases that could be caused by the search space, simulation scenarios with different large search spaces are performed.

Table 3.8 Survey of publications on PID-based controllers and their tuning method for the AVR system

Ref.	Controller	Proposed optimizer	Compared to
[38]	PID	Particle swarm optimization (PSO)	GA
[39]	PID	Continuous action reinforcement learning automata	[38]
[40]	PID	Real-coded GA	[38]
[41]	PID	Bat algorithm (BAT)	[40]
[42]	PID	PSO	[38]
[43]	PID	PSO	∅
[44]	PID	Artificial bee colony (ABC)	PSO and differential evolution (DE)
[45]	PID	Simplified PSO	[44]
[46]	PID	Pattern search (PS)	[44]
[47]	PID	Teaching–learning-based optimization (TLBO)	∅
[48]	PID	Bacterial foraging algorithm (BFA)	Ziegler–Nichols (ZN) and PSO
[49]	PID	Tabu search (TS)	∅
[50]	PID	New third-order PSO	PSO
[51]	PID	Hybrid GA-PSO	PSO and GA
[52]	PID	Taguchi combined genetic algorithm method	PSO and GA
[53]	PID	Alternative PSO	GA and PSO
[54]	PID	Hybrid GA-BFA	GA, PSO, and GA-PSO
[55]	PID	Extended discrete action, Reinforcement learning automata (DARLA)	ZN, DARLA, and GA
[56]	PID	Anarchic society optimization	Craziness-based PSO (CRPSO)
[46]	PID	Many liaison optimization (MOL)	PSO
[57]	PID	Particle swarm optimization with velocity update relaxation	CRPSO and GA [58]
[59]	PID	Chaotic optimization based on Lozi map	∅
[60]	PID	Quantum Gaussian particle swarm optimization (G-QPSO)	PSO and QPSO
[61]	PID	Multi-objective non-dominated sorting genetic algorithm-II (NSGA-II)	∅
[62]	PID	Chaotic Ant Swarm	GA
[63]	PID	Local unimodal sampling (LUS) algorithm	
[64]	PID	TLBO	[44, 45, 62, 63]
[65]	PID	Biogeography-based optimization	[44]
[66]	PID	Symbiotic organisms search and SA	[38, 44, 63–65]
[67]	PID	FIFA world cup competitions	GA, PSO, and imperialist competitive algorithm
[68]	Fuzzy PID	Real-coded GA	GA and LQR
[69]	Fuzzy PID	GA with radial basis function neural network	[68]
[58]	Fuzzy PID	CRPSO	Hybrid Taguchi-PSO, Hybrid-PSO1, Hybrid-PSO2, and GA [38, 70]

(Continued)

Table 3.8 (Continued)

Ref.	Controller	Proposed optimizer	Compared to
	Fractional-order-based PID	GA	PID-based controllers
[72]	FOPID	NSGA-II augmented with a chaotic Henon map	NSGA-II, modified NSGA-II, and PID
[73]	FOPID	NSGA-II	PID
[74]	FOPID	PSO	[75]
[75]	FOPID	Chaotic ant swarm	PSO, GA, ZA, and PID
[37]	FOPID	PSO	PID
[76]	FOPID	ABC with cyclic exchange neighborhood with chaos	PID
[77]	Fuzzy P + Fuzzy I + Fuzzy D	GA-PSO	Fuzzy PID and PID
[78]	PID-Acceleration (PIDA)	TLBO, harmony search algorithm (HSA), LUS	MOL, gravitational search algorithm (GSA), ABC, PSO, DE, and BAT
[79]	PIDA	Current search	TS and GA
[80]	PIDA	HSA	[79]
[81]	PIDA	BAT	[79]
[82]	PID, PIDA	Whale optimization algorithm	[44, 45, 78, 81]

Section 3.2.1 presents the AVR model with the PIDA, $FOPID_f$, and 2DOFPID$_f$ controller structures. Section 3.2.2 provides the simulation results followed by a discussion in Sect. 3.2.3.

3.2.1 Model of the AVR System and its Controllers

The AVR system, which consists of an amplifier, exciter, generator, and sensor, maintains the generator's voltage terminal at a desired value [38]. Figure 3.11 presents the transfer functions of its components [38], and Table 3.9 presents the AVR system variables and constants used in this study. The voltage error $e(s)$, the difference between the generator output voltage $V_t(s)$ and the reference voltage $V_{ref}(s)$, feeds the control system. The control system takes appropriate actions to generate $V_u(s)$ to bring $e(s)$ to zero.

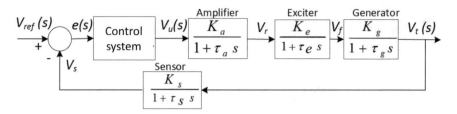

Fig. 3.11 AVR system model

Table 3.9 AVR system variables and constants

Variable	Definition	Value
K_a	Amplifier gain	10
K_e	Exciter gain	1.0
K_g	Generator gain	1.0
K_s	Sensor gain	1.0
τ_a	Amplifier time constant	0.1
τ_e	Exciter time constant	0.4
τ_g	Generator time constant	1.0
τ_s	Sensor time constant	0.01

PIDA Controller

PIDA controller was firstly designed for a third-order system to satisfy transient and steady-state response specifications [84]. The following equation is the transfer function of a PIDA controller [80]:

$$\frac{V_u(s)}{e(s)} = \frac{K_{acc}s^3 + K_{d_1}s^2 + K_{p_1}s + K_{i_1}}{s^3 + \alpha_1 s^2 + \beta_1 s} \tag{3.27}$$

where K_{acc} is the acceleration gain, K_{d_1} the derivative gain, K_{p_1} the proportional gain, K_{i_1} the integral gain, and α_1 and β_1 are filter parameters. The decision variable to optimize is, therefore, $x_{PIDA} = [K_{acc}\ K_{d_1}\ K_{p_1}\ K_{i_1}\ \alpha_1\ \beta_1]$.

FOPID$_f$ Controller

FOPID$_f$ controller possesses two additional terms compared to the original PID$_f$. Theses extra parameters on the derivative and integral actions provide a greater range of freedom, which can increase system stability and robustness [85]. The FOPID$_f$ is expressed in the Laplace domain as follows:

$$FOPID_f(s) = K_{p_2} + K_{i_2}s^{-\lambda} + K_{d_2}\frac{N_2}{1 + N_2\frac{1}{s^\psi}} \tag{3.28}$$

where K_{p_2} is the proportional gain, K_{d_2} the derivative gain, K_{i_2} the integral gain, N_2 the filter coefficient, and λ and ψ are the fractional gains on the integral and derivative actions, respectively. The decision variable is $x_{FOPID_f} = [K_{p_2}\ K_{d_2}\ K_{i_2}\ N_2\ \lambda\ \psi]$.

2DOFPID$_f$ Controller

2DOFPID$_f$ controller has two extra parameters that weight the reference signal before passing through the proportional and derivative actions. This controller is useful to reject disturbances without significantly increasing response overshoot and to dampen the effect of reference signal changes on the control signal. The control signal is expressed as follows:

$$V_u(s) = K_{p_3}\left[p_1 V_{ref}(s) - V_t(s)\right] + \frac{K_{i_3}}{s}e(s) + K_{d_3}\left[p_2 V_{ref}(s) - V_t(s)\right]\frac{N_3}{1 + N\dfrac{1}{s}}$$

(3.29)

where K_{p_3} is the proportional gain, K_{d_3} the derivative gain, K_{i_3} the integral gain, N_3 the filter coefficient, and p_1 and p_2 are the weights on the reference signal before passing through the proportional and derivative actions, respectively. Thus, the decision variable is $x_{2DOFPID_f} = [K_{p_3}\ K_{d_3}\ K_{i_3}\ N_3\ p_1\ p_2]$.

Cost Function

Several cost functions have been proposed for the AVR system control tuning to improve system performances. In [27], an exhaustive literature review highlights the superiority of the cost function proposed in [38]. The cost function in [38] allows optimization algorithms to reach the best dynamic performance trade-off. The cost function used in this study is, therefore, the one proposed in [38] and is as follows:

$$f_{AVR}(x) = \left(1 - e^{-\beta_2}\right)\left(V_{t_{max}} + E_{ss}\right) + e^{-\beta_2}(t_s - t_r),$$

(3.30)

where x is the controller parameter set, β_2 is a weighting factor, $V_{t_{max}}$ is the system response maximum peak value, E_{ss} is the steady-state error, t_s the settling time, and t_r the rising time. For this study, t_s is set to 2% of the reference value, t_r to 10% to 90% of the steady-state value, and $\beta_2 = 1$.

3.2.2   Simulation Results

For all simulations in this subsection, SA and GA paramets are the same as **I-C** and **II-F**, respectively, because the algorithms perform best with these parametrizations—see Table 3.3. Even though these parametrizations might not be optimal for this system, constant parametrization over the different simulations makes the comparison fair. Simulation scenarios are performed with SA and GA for different search space bounds, as presented in Table 3.10.

Table 3.10 Search space intervals

Controllers			Bounds		
PIDA	FOPID$_f$	2DOFPID$_f$	A	B	C
K_{acc}	K_{p_2}	K_{p_3}	[0.1 1,000]	[0.1 500]	[500 1,000]
K_{d_1}	K_{i_2}	K_{i_3}	[0.1 1,000]	[0.1 500]	[500 1,000]
K_{p_1}	K_{d_2}	K_{d_3}	[0.1 1,000]	[0.1 500]	[500 1,000]
K_{i_1}	N_2	N_3	[0.1 1,000]	[0.1 500]	[500 1,000]
α_1	λ	p_1	[0.1 1,000]	[0.1 500]	[500 1,000]
β_1	ψ	p_2	[0.1 1,000]	[0.1 500]	[500 1,000]

Table 3.11 Simulation results obtained with SA and GA algorithms

		SA			GA		
Controllers	Bounds	ME	SD	$f-evals$	ME	SD	$f-evals$
PIDA	A	0.6948	0.0249	8,978	0.6873	0.0030	8,910
	B	0.7053	0.0394	6,724	0.6891	0.0017	9,095
	C	1.7157	0.4191	8,350	1.2515	0.0008	8,000
FOPID$_f$	A	–	–	–	0.8187	0.0070	8,970
	B	–	–	–	0.8206	0.0086	8,990
	C	–	–	–	–	–	–
2DOFPID$_f$	A	28,000	37,947	5,487	0.7822	0.0874	9,005
	B	9,035	3,051	4,221	0.8128	0	9,370
	C	–	–	–	–	–	–

Table 3.11 presents the results for ten simulation runs for each controller and bound intervals. The notation "−" means that the algorithm reached no viable solutions for all runs.

According to ME values, the PIDA controller offers better performances than the other controllers regardless of the optimization algorithm and bounds. Moreover, it is the only controller that could provide feasible solutions for all bounds. Based on ME and SD values, GA provides better performance than SA for all bounds of PIDA. However, performances are close with bound A. For FOPID$_f$, the SA algorithm could not provide any feasible solutions, while GA could reach viable solutions for bounds A and B. Suppose that only SA algorithm was applied to optimize the FOPID$_f$. It would have been tempting to conclude that the FOPID$_f$ controller is not suitable to control the system under study, which is not the case since GA could provide viable solutions. This brings out the relevance of using more than one optimization algorithm for controller tuning when no prior information is known about the controller's suitability for a particular system. If all algorithms cannot provide viable solutions, it most likely means that the controller is not apt for the system. Moreover, this suggests that the SA algorithm is not suitable for solving this optimization problem when no "purposefully chosen" initial state condition is provided—let us recall that initial conditions for SA and the first population for GA are generated randomly. Besides, the simulation runs with the FOPID$_f$ controller highlights the importance and impact of bound intervals. Indeed, while bound C probably does not contain any practicable solutions for the FOPID$_f$ controller, it does not mean that this controller is not suitable for the problem at hand. It is always important to try different search spaces to avoid hasty and wrong conclusions. Along the same line, no viable solutions within bound C for the 2DOFPID$_f$ controller were found by as SA and GA; however, good solutions were found for bounds A and B with GA. Table 3.12 presents the best controller parameters found by SA and GA for each controller. The bounds in which the best solutions were found are also presented.

The optimized PIDAs have the lowest $f_{AVR}(x)$ value, meaning that the PIDA controller offers the best control performance compared to the other controllers. Even if SA has significantly higher ME and SD values than GA for the PIDA controller as presented in Table 3.11, the lowest $f_{AVR}(x)$ value is obtained with the SA algorithm. However, $f_{AVR}(x)$ obtained with GA is very close to the value obtained with SA. Figure 3.12 presents the AVR system responses obtained with the optimized PIDAs.

The system response obtained with the SA-PIDA controller is slightly faster and closer to the reference value during the first 4 s of simulation. GA-PIDA has a small undershoot, but the system response is still within the 2% range of the reference value. Although SA has reached the lowest $f_{AVR}(x)$ for the PIDA controller, in general, GA offers greater optimization performances. As presented in Table 3.11, ME and SD values are lower for all scenarios and bounds.

Figure 3.13 compares the AVR system responses obtained with the parameter sets presented in Table 3.12, i.e., the parameter sets optimized with GA that have reached the best $f_{AVR}(x)$ value for PIDA, FOPID$_f$, and 2DOFPID$_f$ controllers.

According to system responses, the PIDA controller offers the best control performance, which is consistent with $f_{AVR}(x)$ values; PIDA reached the lowest value. Regarding the FOPID$_f$ controller, the system response is coherent with its objective function value $f_{AVR}(x)$. The slightly higher $f_{AVR}(x)$ value compared to PIDA is explained by a higher peak overshoot. On a different note, the system response obtained with 2DOFPID$_f$ could indicate a flaw in the optimization problem setup or objective function. Indeed, despite $f_{AVR}(x)$ value is very close to the value obtained with PIDA, the system response is of poor quality. The $f_{AVR}(x)$ value is low because $t_s - t_r$ is small, the overshoot is small, and there is no steady-state error, which are the three components the objective function penalizes. The objective function is not designed to catch oscillations in the system response unless this behavior is detected in one of the objective function components. A way to address this issue could be by adding t_r or t_s constraints to the optimization problem. However, GA could reach satisfactory system responses for 2DOFPIF$_f$ as Fig. 3.14 presents the second-best response among the 10 runs, for which $f_{AVR}(x) = 0.7190$.

The system response reaches the reference fast without overshoot.

Table 3.12 Best simulation results obtained with SA and GA algorithms for each controller

PIDA	Algorithm	Bound	K_{acc}	K_{d_1}	K_{p_1}	K_{i_1}	α	β	$f_{AVR}(\boldsymbol{x})$
	GA	A	78.88	412.27	1,000	472.70	227.88	946.23	0.6812
	SA	A	33.1424	342.1740	905.7003	678.3521	65.4891	882.4920	0.6788
FOPID$_f$	Algorithm	Bound	K_{p_2}	K_{i_2}	K_{d_2}	N_2	λ	ψ	$f_{AVR}(\boldsymbol{x})$
	GA	B	7.8988	292.0336	7.5109	0.1000	480.6571	1.5220	0.8062
	SA	–	–	–	–	–	–	–	–
2DOFPID$_f$	Algorithm	Bound	K_{p_3}	K_{i_3}	K_{d_3}	N_3	p_1	p_2	$f_{AVR}(\boldsymbol{x})$
	GA	A	2.3370	5.1512	0.4039	999.9067	0.1177	1.0887	0.6898
	SA	B	5.8020	74.3806	1.4227	401.3703	179.9065	412.9024	351.29

Fig. 3.12 AVR system
responses obtained with
optimized PIDA
controllers—SA vs GA

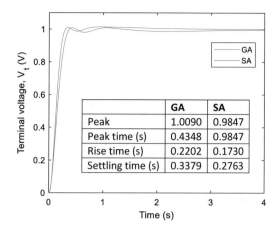

Fig. 3.13 AVR system
responses obtained with GA
optimized controllers

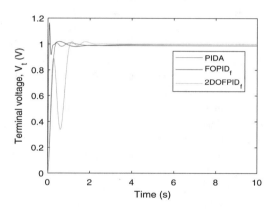

Fig. 3.14 AVR system
responses obtained with GA
optimized controllers

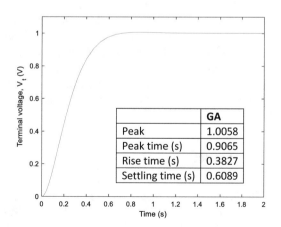

3.2.3   Discussion

The literature survey of PID-based controllers' publications for the AVR system indicated that a limited number of works compare control performances of more complex PID-based controllers. Since metaheuristics are problem-independent, they are an excellent tuning tool to ensure unbiased comparison results. Therefore, this section has proposed the first comparison between PIDA, FOPID$_f$, and 2DOFPID$_f$ controllers tuned by metaheuristics. The controllers had an equal number of parameters to tune for a fair comparison. Also, for more accurate controller comparison, two optimization algorithms, GA and SA, were employed. SA and GA performances were compared, but using two different optimization algorithms limits biases that optimization could introduce.

The GA and SA results show that PIDA controller provides better control performance according to the objective function. Overall, GA offers better optimization performance than SA for all controller tuning. In particular, SA could not provide any viable solutions for FOPID$_f$ controller and could only provide low-quality solutions for 2DOFPID$_f$, while GA could provide satisfactory solutions. As presented in the previous section, the initial point of search influence SA performance. Suppose the starting point is too far away from the optimum region, and the objective function is non-convex with multiple local minimum regions. SA algorithm might get trapped in one of the local minimum regions. Furthermore, the simulation results strengthen the relevance of using two optimization algorithms when comparing controller performance. SA results suggest that FOPID$_f$ controller is not suitable for the system under study since no viable solutions have been found. However, GA could locate satisfactory results, which indicates that the FOPID$_f$ controller is appropriate.

Besides, the importance of search space has been investigated. GA and SA have optimized each controller for different search spaces. No viable solutions could be found for some search spaces. Thus, trying more than one search space when no prior information is known about potential controller parameter values is necessary. Indeed, GA could not provide any feasible solution for FOPID$_f$ with bound C but could for bounds A and B. Thus, suppose bound C would have been the only bound used for the optimization of the FOPID$_f$ controller. Wrong conclusions would probably have been drawn, i.e., FOPID$_f$ is not suitable for the AVR system.

3.3   Conclusion

Metaheuristics applied to control tuning have been studied on two canonical benchmark systems in control theory: the ICP and AVR systems. The main advantages of metaheuristics for controller tuning and the major elements that impact algorithm performance have been demonstrated.

Principal Features of Metaheuristics for Controller Tuning

- Problem-independent—Metaheuristics are appropriate for linear and non-linear systems.
- Controller-independent—Optimization algorithms are suitable for several controller structures from conventional control to intelligent control.
- Application-customized objective function—Using metaheuristics for control tuning allows targeting system requirements during the optimization along with considering multiple system constraints.
- Robustness and stability properties—It is possible to include robustness and stability criteria within the optimization process.
- Controller tuning tool for comparison—Given the advantages mentioned above, metaheuristics are great tuning methods to compare performances of different controllers.

Major Elements That Influence Metaheuristics Performances when Applied to Controller Tuning

- Algorithm parametrization—It is recommended to employ a metaheuristic with a parametrization guide to help set the algorithm parameters. If not, algorithm parameter default values are a reasonable choice [34], and making some modifications from default values is an effective way to explore the possibility of better parametrization.
- Stop criteria—The optimization process is optimized itself, i.e., getting a satisfactory solution as fast as possible. Therefore, the stop criterion is an important element that influences performance. To ensure higher computational costs are worth it, the response improvement should be evaluated.
- Search space—As demonstrated, optimization search space plays a crucial role in algorithm performance since it imposes where the algorithm searches for solutions. For controller tuning problems for which no prior knowledge of the possible parameter values is known, testing different search spaces is advisable. Moreover, the search space influences the convergence speed of the algorithm; most likely, the bigger the search space is, the slower the algorithm converges.

(continued)

- Cost function—Cost function should reflect as much as possible the system specifications. A cost function designed inadequately for the system provides inadequate solutions. Therefore, control designers must be careful, as unsatisfactory solutions may be obtained due to the cost function and not due to the algorithm or controller.

Moreover, SA and GA performances for control tuning have been compared. Overall, GA provides better performance than SA for the systems under study, which substantiates the concept that population-based algorithms offer better performance than trajectory-based algorithms. The SA and GA performance comparison has followed ethical practices for algorithm benchmarking. We have provided the most important information to report the results fairly. More information regarding optimization algorithm benchmarking is presented in [86–88].

Components for Proper Algorithm Benchmarking

- Provide the algorithms used or provide references to obtain them
- Mention the algorithm parametrization employed
- Describe the computational environment
- Employ the same cost function, search space, and stop criteria
- Use different metrics to assess algorithm performance.

The following chapter presents open questions, critical challenges, and future research trends in the field of metaheuristics.

References

1. Gu, Y.: Time-dependent nonlinear control of bipedal robotic walking. Dissertation Purdue University (2017)
2. Fantoni, I., Lozano, R.: Non-linear Control for Underactuated Mechanical Systems, Springer, New York (2001)
3. Gu, Y., Yao, B., Lee, G.: Exponential stabilization of fully actuated planar bipedal robotic walking with global position tracking capabilities. J. Dyn. Syst. Meas. Control **140**(5), 051008 (2018)
4. Duan, M., Okwudire, C.E.: Proxy-based optimal control allocation for dual-input over-actuated systems. IEEE/ASME Trans. Mechatron. **23**, 895–905 (2018)
5. Boubaker, O.: The inverted pendulum benchmark in nonlinear control theory: a survey. Int. J. Adv. Robot. Syst. **10** (2013)

6. Blondin, M.J., Pardalos, P.M.: A holistic optimization approach for inverted cart-pendulum control tuning. Soft Comput. **24**, 4343–4359 (2020)
7. Aström, K.J., Furuta, K.: Swinging up a pendulum by energy control. Automatica **36**(2), 287–295 (2000)
8. Kumar, E.V., Jerome, J.: Robust LQR controller design for stabilizing and trajectory tracking of inverted pendulum. Proc. Eng. **64**, 169–178 (2013)
9. Blondin, M.J., Sicard, P., Pardalos, P.M.: The ACO-NM algorithm for controller tuning for an inverted cart-pendulum. In: 2018 International Symposium on Power Electronics, Electrical Drives, Automation and Motion (SPEEDAM), pp. 1370–1375. IEEE, Piscataway (2018)
10. Chatterjee, D., Patra, A., Joglekar, K.H.: Swing-up and stabilization of a cart–pendulum system under restricted cart track length. Syst. Control Lett. **47**(4), 355–364 (2002)
11. Bradshaw, A., Shao, J.: Swing-up control of inverted pendulum systems. Robotica **14**(4), 397–405 (1996)
12. Nundrakwang, S., Benjanarasuth, T., Ngamwiwit, J., Komine, N.: Hybrid controller for swinging up inverted pendulum system. In: 2005 5th International Conference on Information Communications Signal Processing, pp. 488–492. IEEE, Piscataway (2005)
13. Muskinja, N., Tovornik, B.: Swinging up and stabilization of a real inverted pendulum. IEEE Trans. Ind. Electron. **53**(2), 631–639 (2006)
14. Oróstica, R., Duarte-Mermoud, M.A., Jáuregui, C.: Stabilization of inverted pendulum using LQR, PID and fractional order PID controllers: a simulated study. In: IEEE International Conference on Automatica, pp. 1–7 (2016)
15. Howimanporn, S., Thanok, S., Chookaew, S., Sootkaneung, W.: Design and implementation of PSO based LQR control for inverted pendulum through PLC. In: IEEE/SICE International Symposium on System Integration, pp. 664–669 (2016)
16. Hassani, K., Lee, W.S.: Multi-objective design of state feedback controllers using reinforced quantum-behaved particle swarm optimization. Appl. Soft Comput. **41**, 66–76 (2016)
17. Wongsathan, C., Sirima, C.: Application of GA to design LQR controller for an inverted pendulum system. In: IEEE International Conference on Robotics and Biomimetics, pp. 951–954 (2008)
18. Almobaied, M., Eksin, I., Guzelkaya, M.: Design of LQR controller with big bang-big crunch optimization algorithm based on time domain criteria. In: Mediterranean Conference on Control and Automation, pp. 1192–1197 (2016)
19. Jacknoon, A., Abido, M.A.: Ant colony based LQR and PID tuned parameters for controlling inverted pendulum. In: International Conference on Communication, Control, Computing and Electronics Engineering (ICCCCEE) (2017)
20. Mua'zu, M.B., Salawudeen, A.T., Sikiru, T.H., Abdu, A.I., Mohammad, A.: Weighted artificial fish swarm algorithm with adaptive behaviour based linear controller design for nonlinear inverted pendulum. J. Eng. Res. **20**(1), 1–12 (2015)
21. Ata, B., Coban, R.: Artificial bee colony algorithm based linear quadratic optimal controller design for a nonlinear inverted pendulum. Int. J. Intell. Syst. Appl. Eng. **3**(1), 1–6 (2015)
22. Ha, K.J., Kim, H.M.: A genetic approach to the attitude control of an inverted pendulum system. In: Proceedings Ninth IEEE International Conference on Tools with Artificial Intelligence, pp. 268–269. IEEE (1997)
23. Henmi, T., Park, Y., Deng, M., Inoue, A.: Stabilization controller for a cart-type inverted pendulum via a partial linearization method. In Proceedings of the 2010 International Conference on Modelling, Identification and Control (pp. 248–253). IEEE, Piscataway (2010)
24. Wang, H., Dong, H., He, L., Shi, Y., Zhang, Y.: Design and simulation of LQR controller with the linear inverted pendulum. In 2010 International Conference on Electrical and Control Engineering, pp. 699–702. IEEE, Piscataway (2010)
25. Ozana, S., Pies, M., Slanina, Z., Hajovsky, R.: Design and implementation of LQR controller for inverted pendulum by use of REX control system. In: 2012 12th International Conference on Control, Automation and Systems, pp. 343–347. IEEE, Piscataway (2012)
26. Prasad, L.B., Tyagi, B., Gupta, H.O.: Optimal control of nonlinear inverted pendulum system using PID controller and LQR: performance analysis without and with disturbance input. Int. J. Autom. Comput. **11**(6), 661–670 (2014)

27. Blondin, M.J., Sanchis, J., Sicard, P., Herrero, J.M.: New optimal controller tuning method for an AVR system using a simplified ant colony optimization with a new constrained Nelder–Mead algorithm. Appl. Soft Comput. **62**, 216–229 (2018)
28. Blondin, M.J., Sicard, P., Pardalos, P.M.: Controller tuning approach with robustness, stability and dynamic criteria for the original AVR system. Math. Comput. Simul. **163**, 168–182 (2019)
29. Quanser Inc.: Quanser Inc. SIP and SPG user manual (2009)
30. Apkarian. J., Lacheray, H., Martin, P.: Linear pendulum gantry experiment for MAT-LAB/simulink users. In: Instructor Workbook. In: Quanser, Markham, pp. 1–40 (2012)
31. Lewis, F.L., Draguna V., Syrmos, V.L.: Optimal Control Wiley, Hoboken (2012)
32. Blasco, X.: Basic genetic algorithm (2020). https://www.mathworks.com/matlabcentral/fileexchange/39021-basic-genetic-algorithm. Retrieved 13 Feb. 2020
33. Blasco Ferragud, F.X.: Control predictivo basado en modelos mediante técnica de optimización heurística. Ph.D. Thesis (en espagnol) Editorial UPV (1999). ISBN 84–699-5429-6.
34. Andrea, A., Fraser, G.: Parameter tuning or default values? An empirical investigation in search-based software engineering. Empirical Softw. Eng. **18**(3), 594–623 (2013)
35. Rodriguez, F.J., Garcia-Martinez, C., Lozano, M.: Hybrid metaheuristics based on evolutionary algorithms and simulated annealing: taxonomy, comparison, and synergy test. IEEE Trans. Evol. Comput. **16**(6), 787–800 (2012)
36. Pan, I., Das, S.: Frequency domain design of fractional order PID controller for AVR system using chaotic multi-objective optimization. Int. J. Electr. Power Energy Syst. **51**, 106–118 (2013)
37. Zamani, M., Karimi-Ghartemani, M., Sadati, N., Parniani, M.: Design of a fractional order PID controller for an AVR using particle swarm optimization. Control Eng. Pract. **17**(12), 1380–1387 (2009)
38. Gaing, Z.-L.: A particle swarm optimization approach for optimum design of PID controller in AVR system. IEEE Trans. Energy Convers. **19**, 384–391 (2004)
39. Kashki, M., Abdel-Magid, Y.L., Abido, M.A.: A reinforcement learning automata optimization approach for optimum tuning of PID controller in AVR system. In: D.-S. Huang et al. (eds.) International Conference on Intelligent Computing, pp. 684–692 (2008)
40. C.-C. Wong, S.-A. Li, H.-Y. Wang, Optimal PID controller design for AVR system. Tamkang J. Sci. Eng. **12**, 259–270 (2009)
41. O. Bendjeghaba, I.B. Saida, Bat algorithm for optimal tuning of PID controller in an AVR system. In: International Conference on Control Engineering and Information Technology, pp. 158–170 (2014)
42. Rahimian, M.S., Raahemifar, K.: Optimal PID controller design for AVR system using particle swarm optimization algorithm. In: Canadian Conference on Electrical and Computer Engineering (2011), pp. 337–340
43. Sahib, M.A.: A New Multi-Objective Performance Criterion used in PD Tuning Optimization Algorithms. J. Adv. Res. **7**(1), 125–34 (2015)
44. Godze, H., Taplamacioglu, M.C.: Comparative performance analysis of artificial bee colony algorithm for automatic voltage regulator (AVR) system. J. Franklin Inst. **348**, 1927–1946 (2011)
45. Panda, S., Sahu, B.K., Mohanty, P.K.: Design and performance analysis of PID controller for an automatic voltage regulator system using simplified particle swarm optimization. J. Franklin Inst. **349**, 2609–2625 (2012)
46. Sahu, B.K., Mohanty, P.K., Panda, S., Kar, S.K., Mishra, N.: Design and comparative performance analysis of PID controlled automatic voltage regulator tuned by many optimizing liaisons. In: International Conference Advances in Power Conversion and Energy Technologies, pp. 1–6 (2012)
47. Priyambada, S., Mohanty, P.K., Sahu, B.K.: Automatic voltage regulator using TLBO algorithm optimized PID controller. In: 2014 9th International Conference on Industrial and Information Systems (ICIIS), pp. 1–6 (2014)
48. Rostami, M., Heydari, G., Gharaveisi, A.A., Rafie, S.M.R.: Optimal design of PID controller using bacterial foraging algorithm for AVR system. In: Congress on Fuzzy and Intelligent System (2008)

49. Oonsivilai, A., Pao-La-Or, P.: Application of adaptive tabu search for optimum PID controller tuning AVR system. WSEAS Trans. Power Syst. **6**, 496–506 (2008)
50. Kang, H., Kwon, M.W., Bae, H.G.: PID coefficient designs for the automatic voltage regulator using a new third-order particle swarm optimization. In: International Conference on Electronics and Information Engineering, pp. 179–183 (2010)
51. Amer, M.L., Hassan, H.H., Hosam, M.Y.: Modified evolutionary particle swarm optimization for AVR-PID tuning. In: Communications and Information Technology Systems and Signals, pp. 164–173 (2008)
52. Hasanien, H.M.: Design optimization of PID controller in automatic voltage regulator system using Taguchi combined genetic algorithm method. IEEE Syst. J. **7**, 825–831 (2013)
53. Shabib, G., Gayed, M., Rashwan, A.M.: Optimal tuning of PID controller for AVR system using modified particle swarm optimization. In: International Middle East Power Systems Conference, pp. 305–310 (2010)
54. Kim, D.H.: Hybrid GA-BF based intelligent PID controller tuning for AVR system. Appl. Soft Comput. **11**, 11–22 (2011)
55. Mohammadi, S.M.A., Gharaveisi, A.A., Mashinchi, M., Rafiei, S.M.R.: New evolutionary methods for optimal design of PID controllers for AVR system. In: IEEE Bucharest Power Tech Conference, pp. 1–8 (2009)
56. Shayeghi, H., Dadashpour, J.: Anarchic society optimization based PID control of an automatic voltage regulator (AVR) system. Electr. Electron. Eng. **2**, 199–207 (2012)
57. Chatterjee, A., Mukherjee, V., Ghoshal, S.P.: Velocity relaxed and craziness-based swarm optimized intelligent PID and PSS controlled AVR system. Electr. Power Energy Syst. **31**, 323–333 (2009)
58. Mukherjee, V., Ghoshal, S.P.: Intelligent particle swarm optimized fuzzy PID controller for AVR system. Electr. Power Syst. Res. **77**, 1689–1698 (2007)
59. dos Santos Coelho, L.: Tuning of PID controller for an automatic regulator voltage system using chaotic optimization approach. Chaos Solutions Fractals **39**, 1504–1514 (2009)
60. dos Santos Coelho, L., de M. Herrera, B.A.: Quantum Gaussian particle swarm optimization approach for PID controller design in AVR system. In: IEEE International Conference on Systems, Man and Cybernetics, pp. 3708–3713 (2008)
61. Yegireddy, N.K., Panda, S.: Design and performance analysis of PID controller for an AVR system using multi-objective non-dominated shorting genetic algorithm-II. In: International Conference on Smart Electric Grid, pp. 1–7 (2014)
62. Zhu, H., Li, L., Zhao, Y., Guo, Y., Yang, Y.: CAS algorithm-based optimum design of PID controller in AVR system. J. Chaos Solitons Fractals **42**, 792–800 (2009)
63. Mohanty, P.K., Sahu, B.K., Panda, S.: Tuning and assessment of proportional–integral–derivative controller for an automatic voltage regulator system employing local unimodal sampling algorithm. Electr. Power Compon. Syst. **42**(9), 959–969 (2014). https://doi.org/10.1080/15325008.2014.903546
64. Chatterjee, S., Mukherjee, V.: PID controller for automatic voltage regulator using teaching–learning based optimization technique. Int. J. Electr. Power Energy Syst. **1**(77), 418–429 (2016)
65. Güvenç, U., Işik AH, Y.T., Akkaya, I.: Performance analysis of biogeography-based optimization for automatic voltage regulator system. Turkish J. Electr. Eng. Comput. Sci. **23**(3), 1150–1162 (2016)
66. Çelik, E., Öztürk, N.: A hybrid symbiotic organisms search and simulated annealing technique applied to efficient design of PID controller for automatic voltage regulator. Soft. Comput. **22**, 8011–8024 (2018). https://doi.org/10.1007/s00500-018-3432-2
67. Razmjooy, N., Khalilpour, M., Ramezani, M.: A new meta-heuristic optimization algorithm inspired by FIFA world cup competitions: theory and its application in PID designing for AVR system. J. Control Autom. Electr. Syst. **27**(4), 419–440 (2016)
68. Devaraj, D., Selvabala, B.: Real-coded genetic algorithm and fuzzy logic approach for real-time tuning of proportional-integral-derivative controller in automatic voltage regulator system. IET Gener. Transm. Distrib. **3**, 641–649 (2008)

69. Al Gizi, A.J.H., Mustafa, M.W., Al-geelani, N.A., Alsaedi, M.A.: Sugeno fuzzy PID tuning, by genetic-neutral for AVR in electrical power generation. Appl. Soft Comput. **28**, 226–236 (2015)
70. Ghoshal, S.P.: Optimizations of PID gains by particle swarm optimizations in fuzzy based automatic generation control. Electr. Power Syst. Res. **72**(3), 203–212 (2004)
71. Camacho, N.A., Duarte-Mermoud, M.A.: Fractional adaptive control for an automatic voltage regulator. ISA Trans. **52**, 807–815 (2013)
72. Pan, I., Das, S.: Frequency domain design of fractional order PID controller for AVR system using chaotic multi-objective optimization. Electr. Power Energy Syst. **51**, 106–118 (2013)
73. Pan, I., Das, S.: Chaotic multi-objective optimization based design of fractional order PIλDμ controller in AVR system. Int. J. Electr. Power Energy Syst. **43**, 1–30 (2012)
74. Ramezanian, H., Balochian, S., Zare, A.: Design of optimal fractional-order PID controllers using particle swarm optimization algorithm for automatic voltage regulator (AVR) system. J. Control Autom. Electr. Syst. **24**, 601–611 (2013)
75. Tang, Y., Cui, M., Hua, C., Li, L., Yang, Y.: Optimum design of fractional order PIλDμ controller for AVR system using chaotic ant swarm. Exp. Syst. Appl. **39**(8), 6887–6896 (2012)
76. D.L. Zhang, Y.C. Tang, X.P.Guan, Optimum design of fractional order PID controller for an AVR system using an improved artificial bee colony algorithm. Acta Autom. Sin. **40**, 973–979 (2014)
77. Shayeghi, H., Younesi, A., Hashemi, Y.: Optimal design of a robust discrete parallel FP + FI + FD controller for the automatic voltage regulator system. Electr. Power Energy Syst. **67**, 66–75 (2015)
78. Mosaad, A.M., Attia, M.A., Abdelaziz, A.Y.: Optimization techniques to tune the PID and PIDA controllers for AVR performance enhancement. i-manager's J. Instrum. Control Eng. **5**(1), 1–10 (2017)
79. Puangdownreong, D.: Application of current search to optimum PIDA controller design. Intell. Control Autom. **3**(4), 303–312 (2012)
80. Sambariya, D.K., Paliwal, D.: Optimal design of PIDA controller using harmony search algorithm for AVR power system. In: 2016 IEEE 6th International Conference on Power Systems (ICPS). IEEE, Piscataway (2016)
81. Sambariya, D.K., Paliwal, D.: Design of PIDA controller using bat algorithm for AVR power system. Adv. Energy Power **4**(1), 1–6 (2016)
82. Mosaad, A.M., Attia, M.A., Abdelaziz, A.Y.: Whale optimization algorithm to tune PID and PIDA controllers on AVR system. Ain Shams Eng. J. **10**(4), 755–767 (2019)
83. Blondin, M., Sicard, P.: A hybrid ACO and Nelder-Mead constrained algorithm for controller and anti-windup tuning. In: 2014 16th European Conference on Power Electronics and Applications, Lappeenranta, pp. 1–10 (2014)
84. Jung, S., Dorf, R.C.: Analytic PIDA controller design technique for a third order system. In: Proceedings of 35th IEEE Conference on Decision and Control, Kobe, vol. 3, pp. 2513–2518 (1996). https://doi.org/10.1109/CDC.1996.573472
85. Mande, D., Blondin, M., Pedro, J., Trovão, F.: Optimization of fractional order PI controller for bidirectional quasi-Z-source inverter used for electric traction system. IET Electr. Syst. Transp. (2020)
86. Barr, R.S., Golden, B.L., Kelly, J.P., Resende, M.G., Stewart, W.R.: Designing and reporting on computational experiments with heuristic methods. J. Heuristics **1**(1), 9–32 (1995)
87. Beiranvand, V., Hare, W., Lucet, Y.: Best practices for comparing optimization algorithms. Optim. Eng. **18**, 815–848 (2017). https://doi.org/10.1007/s11081--017-9366-1
88. Rardin, R.L., Uzsoy, R.: Experimental evaluation of heuristic optimization algorithms: a tutorial. J. Heuristics **7**(3), 261–304 (2001)
89. Bingul, Z., Karahan, O.: A novel performance criterion approach to optimum design of PID controller using cuckoo search algorithm for AVR system. J. Franklin Inst. **355**(13), 5534–5559 (2018)
90. Priyambada, S., Mohanty, P.K., Sahu, B.K.: Automatic voltage regulator using TLBO algorithm optimized PID controller. In 2014 9th International Conference on Industrial and Information Systems (ICIIS), pp. 1–6. IEEE, Piscataway (2014)

91. Sahu, B.K., Mohanty, P.K., Panda, S., Mishra, N.: Robust analysis and design of PID controlled AVR system using pattern search algorithm. In: IEEE International Conference on Power Electronics, Drives and Energy Systems, pp. 1–6 (2012)
92. Brown, B., Aaron, M.: The politics of nature. In: Smith, J. (ed.) The Rise of Modern Genomics, 3rd edn. Wiley, New York (2001)
93. Dod, J.: Effective substances. In: The Dictionary of Substances and Their Effects. Royal Society of Chemistry (1999). Available via DIALOG. http://www.rsc.org/dose/titleofsubordinatedocument. Retrieved 15 Jan. 1999
94. Slifka, M.K., Whitton, J.L.: Clinical implications of dysregulated cytokine production. J. Mol. Med. **78**, 74–80 (2000). https://doi.org/10.1007/s001090000086
95. Smith, J., Jones, M. Jr., Houghton, L. et al.: Future of health insurance. N. Engl. J. Med. **965**, 325–329 (1999)
96. South, J., Blass, B.: The Future of modern Genomics. Blackwell, London (2001)

Chapter 4
Future Direction and Research Trends

4.1 Algorithm Contributions and Assessments

A new algorithm is relevant to the scientific community if it brings contributions. There exist common standards to assess optimization algorithm contributions, which are [1]:

1. computation speed—the algorithm reaches the desired response faster than other strategies,
2. precision—it achieves a better solution based on the objective function value than other techniques,
3. robust—performance is less influenced by the algorithm and problem characteristics than other methods,
4. simple—the algorithm is easy to implement,
5. high impact—it solves a relevant or new optimization problem quicker and more precisely than other algorithms,
6. generalizable—the algorithm is suitable for a large spectrum of problems.

The widespread practice is to demonstrate algorithm contributions through computational experimentation. Computational investigations typically consist of comparing the new algorithm to other algorithms on benchmark functions and/or benchmark engineering problems. Since there is no standard comparison benchmark and protocol established yet, the designer creates their computational testing by Barr et al. [1]:

1. selecting the engineering problems or test functions to be solved,
2. choosing the algorithms to be compared with,
3. deciding the computing environment,
4. coding the algorithms or using already coded algorithms,

M. J. Blondin, *Controller Tuning Optimization Methods for Multi-Constraints and Nonlinear Systems*, SpringerBriefs in Optimization, https://doi.org/10.1007/978-3-030-64541-0_4

5. choosing the performance measures,
6. setting algorithm parametrizations,
7. reporting the results.

While this comparison approach seems a natural and effective way of demonstrating algorithm performance and superiority, it can be problematic for several reasons.

As the designer has free rein to build their benchmark, it can be done to prove their point, which can lead to unethical approach and biased conclusions [1, 2]. As demonstrated in the previous chapter, any decisions made by the designer impact the final results. Therefore, customized benchmarks open the door to self-serving choices regarding algorithm parametrization, benchmark functions, and algorithms used in the benchmark test to demonstrate the proposed algorithm's superiority.

Secondly, even with the most care to have unbiased and authentic comparisons, conclusions made concerning the algorithm performance are true for that specific comparison with that particular environment setup. For example, algorithm X may perform better than algorithm Y. However, if the search space is changed, algorithm Y may perform better than algorithm X. Consequently, the literature abounds with publications presenting results of different benchmark algorithm comparisons, and sometimes the conclusions seem to contradict one another. As presented in the survey in Table 2.1, this myriad of different benchmark comparisons complicates the literature review to pinpoint the most suitable optimization algorithm for the problem at hand. Simultaneously, algorithm comparisons should be made with the current state-of-the-art and the best algorithms [3]. However, it becomes hard to identify them because benchmarks are different, and the current-state-of-the-art is diluted with a high number of publications.

Another element that plays against appropriate comparison is the diversity of algorithm proposals. As presented in Chap. 2, hundreds of algorithms. exist, making it impossible to compare a new proposal with all algorithms. A careful designer may want to select the most meaningful algorithms to compare their proposal to have an accurate and unbiased comparison as much as possible. However, many algorithms do not present clearly their exclusive advantages/benefits from a theoretical and mathematical perspective. Therefore, it becomes challenging to create a meaningful and fair computational benchmark.

4.2 Recommendation for Future Research

Attempts have been made to overcome the disadvantage of customized benchmarks. Conferences have been proposing competitions to solve different types of optimization problems. For instance, the 2020 Genetic and Evolutionary Computation Conference runs a competition containing 11 types of optimization problems such as Single-Objective Bound Constrained Numerical Optimization and Evolutionary Multi-task Optimization. In the same vein, the 2020 IEEE World Congress on Computational Intelligence has a contest consisting of 14 categories of optimization problems. The conferences rely on experts in the optimization-related field to create meaningful optimization problems. The competitors download the optimization problem codes on the competition websites as well as the documentation regarding the rules and criteria to follow. These competitions give an authentic platform to algorithm designers to compare their algorithms against other recent proposals. The possibilities of unethical approach and unbiased results are minimized, considering that the algorithm designers have not created the benchmarks. Moreover, since the competition results are published in conference papers, any new algorithm could compare its performance to the best competition results and state-of-the-art algorithms. Optimization algorithm competitions seem to be promising to address critical issues in algorithm comparisons. However, these standard benchmarks are not well established in the community. Indeed, most of the new algorithms' publications still compare algorithm performances on customized benchmarks with well-known algorithms such as GA and PSO [3]. A solution to this matter would be establishing an armamentarium of programs and protocols designed for algorithm comparisons [2]. In addition, enforcing the use of more standard benchmarks and state-of-the-art algorithms by journal and conference reviewers and editorial boards when dealing with a paper that proposes a new algorithm would be beneficial for this research field.

Nonetheless, standard benchmarks are platforms for comparing algorithms to keep track of the best algorithm to date. The algorithm achieving the best performance wins. Indeed, the mainstream of research is in the up-of- the-hill race of the best algorithms, without any fundamental comprehension of algorithms, i.e., if the algorithm provides better results than others, it is enough to be published. While performance is important in optimization, other features matter. In particular, theoretical and mathematical understanding of algorithms are important issues in the domain. Therefore, authentic algorithms' comparison either through conference competitions or comparison programs and protocols is only a part of the solution to the field's issues. Research gaps between algorithm performance and fundamental reasons for such performance are to be filled.

The publication of new optimization algorithms has skyrocketed over recent years. While diversity usually brings a positive impact on a field of research, this huge number of proposals may be damaging. Many recent proposed algorithms, mostly metaphor-based, prove their efficiency exclusively through customized computational benchmarks, i.e., with limited mathematical or theoretical analyses to support their advantages/superiority. Also, scientific sets of guidelines for their application are sometimes lacking. Thus, diligently selecting the most appropriate algorithm to solve the problem at hand becomes challenging even for knowledgeable researchers. Therefore, the research community is encouraged to bridge the gap between empirical performance assessments and a thorough mathematical understanding of algorithm properties [4]. Rigorous mathematical understanding includes theoretical and mathematical studies on metaheuristics properties such as proof of convergence, convergence rate, and computational complexity. Indeed, only a limited number of algorithms cover certain convergence properties. In [5], based on Markov's theory, theoretical results regarding the convergence of evolutionary algorithms are presented, but the practical estimation has not been developed yet. Along the same line, convergence rates of GA are analyzed in [6]. Some convergence proofs for ACO are available in [7–9]. Along the same line, uncovering why specific algorithms have good performances on particular problems deserves more attention. Indeed, demonstrating the connection between algorithm properties and problem structures will provide facilitate the selection and application of optimization algorithms to new problems.

Besides, metaheuristics possess many internal parameters that lead their search. Time should be spent on assigning appropriate values to these parameters since algorithm parametrization significantly influences algorithm performance, as demonstrated in Chap. 3. The typical practice is off-line tuning, i.e., the designer sets the parameters before running the algorithm, which remains unaltered during the entire optimization process. While default values could be a reasonable choice [10], usually, the optimization designer try several parameter values to achieve a (near)-optimal solution. However, the aim is not achieving the best algorithm parametrization possible, as it would be solving another optimization problem itself, which would be very time-consuming. Some tools, such as REVAC, automate the process of algorithm parameter tuning and return the best parametrization. While it alleviates researchers from parameter tuning and provides an optimized metaheuristic, these options have some drawbacks. Indeed, even if this process is automated, the tuning task itself remains time-consuming. Moreover, the tuned parametrization is to be used for off-line optimization. The tuned metaheuristic becomes problem-dependent, i.e., the optimized parametrization is valid for a particular problem. Therefore, open challenges remain, i.e., having a proper parametrization for online optimization and keeping the metaheuristic problem-independent.

Attempts have been made to provide some guidelines to facilitate algorithm parametrization when algorithms are applied to new problems [11–13]. The guidelines are based either on the number of variables the algorithm has to optimize or on the type of functions such as convex, non-convex, and multi-modal. Another current research direction is to develop smart algorithms that tune their parameters. For instance, a strategy called self-adaptive is implemented within the algorithm and modifies the parameters during the optimization process. Self-adaptive strategies make the parameters evolve according to the results obtained during the optimization process [4]. Another research direction is to implement mechanisms exclusively on parameters that strongly influence algorithm performance. For example, current trends of research are proposing self-adaptive population mechanisms for population-based algorithms [14]. Population size is one of the most important parameters that influences the exploration of the search space. A more significant population brings more variety in potential solutions, which is valuable at the beginning of the exploration process, but may slow down the intensification process toward the end of the optimization. However, self-adaptive parametrization approaches are often based on statistical results of several runs of the optimization algorithm [15] and not necessarily on a deep understanding of the parameter effects from a mathematical perspective. The ultimate goal is developing smart optimization algorithms, i.e., algorithms auto-tune their parameters based on instant feedback received during the optimization process. Along the same line, assessing and gauging the diversification versus intensification automatically while the algorithm is running is an open area that deserves more research [16].

4.3 Hybrid Metaheuristics

The keen interest in applying metaheuristics to various optimization problems has spotlighted strengths and weaknesses that metaheuristics may have in solving some practical problems. To overcome the weaknesses and leverage metaheuristics's strengths, an effective research direction is combining different algorithms [17]. These combined algorithms are called hybrid algorithms, in which algorithms complement each other. A widespread hybridization is combining population-based algorithms with local optimization algorithms or trajectory-based algorithms [18]. Population-based algorithms are powerful methods to explore the search space and target promising areas. In contrast, local search algorithms or trajectory-based algorithms effectively intensify the search in the promising areas to quickly reach (near)-optimal solutions. For instance, the SA algorithm is an excellent tool for optimization, as shown in the previous chapter, but its performance depends significantly on the initial decision variables. This dependency makes SA algorithm a perfect candidate for hybridization. SA is used as a local search to find better solutions in the neighborhood of promising regions found by another algorithm. Indeed, the SA algorithm has been combined with many other algorithms, such as Whale Optimization Algorithm [19], Particle Swarm Optimization [20], GA [21], Symbiotic Organism Search [22], and Ant Colony system [23]. It has been proven that these combinations enhance algorithm performance compared to those that were employed separately. Also, algorithms from different optimization classes can be combined, e.g., a deterministic optimization algorithm and a metaheuristic approach. Indeed, combining algorithms may improve performance by either reaching better solutions or reaching satisfactory solutions faster. However, many publications demonstrate the superiority of their hybrid algorithm propositions by comparing it with other algorithms. Therefore, in the same line of thoughts as in the previous section, theoretical studies to provide a mathematical understanding of hybrid algorithms behavior and convergence should be more investigated.

4.3.1 Hybrid Algorithms: Controller Tuning

Some hybrid algorithms have been proposed specifically for controller tuning such as Genetic Algorithms-Bacteria Foraging [24], Genetic Algorithm-Gravitational Search Algorithm [25], Artificial Bee Colony Algorithm with the sequential quadratic programming [26], and Ant Colony Optimization algorithm combined to Nelder–Mead (ACO-NM) [27]. In particular, the ACO-NM has been employed for many applications, such as the inverted cart pendulum [28], cruise control for an electric vehicle [29], and fractional order proportional-integral controllers for a bidirectional quasi-Z-source inverter for an electric vehicle [30]. The ACO-NM in [27] proposed a simplified version of ACO to reduce the number of ACO parameters to tune. Moreover, since the original NM method is for unconstrained problems, a new procedure to constrain NM is proposed. It is important to employ an algorithm that can deal with controller constraints as some parameters must be limited to ensure system stability and robustness or to respect system requirements. There exist different approaches to constrain NM method such as positivity constraint [31], near-value limitations [32], and reflection constraint [33]. While these approaches have demonstrated some degree of effectiveness, the proposed algorithm to constrain NM in [27] is tailored for controller tuning and has shown superior performance. Concretely, the other procedures deal only with the parameters that exceed the bounds. Suppose a parameter exceeds its bound. The constraint algorithm will change only the value of this parameter; the other terms remain unchanged. While that ensures the bounds are respected, these procedures do not consider the interaction between controller parameters. Considering parameter interaction in control tuning is very intuitive. Indeed, in manual control tuning, a common practice is to set one parameter at a time because the value of one parameter somehow affects the value of the next parameter. Therefore, to emulate the reasoning behind manual control tuning and leverage its advantages when dealing with constraints, the proposed procedure to constrain NM [27] considers the interaction between controller parameters. In particular, if a parameter steps out of its bounds, the ratio of the excess value over the distance between the value outside bound and the value to be replaced in the NM simplex is computed. Thereafter, the entire simplex in the NM method is shrunk by this ratio. If there is more than one value out of bounds, the highest ratio is taken.

To demonstrate the efficiency of hybrid algorithms, the ACO-NM has been applied to tune the controllers for the AVR system presented in Chap. 3. The simulation environment is the same as Sect. 3.2.2. The ACO-NM parametrization is set to its default values, as presented in [27]. The ACO-NM is run ten times. The best solution reached by ACO-NM for each controller is presented in Table 4.1 and compared to the best solutions reached by GA and SA, previously presented in Table 3.12. *nbEvals* refers to the number of function evaluations needed for the algorithm to reach the best solution. The best results are in bold in Table 4.1.

The results show that the ACO-NM reaches better solutions than GA and SA for all controllers. In particular, the number of function evaluations is significantly lower for PIDA and 2DOFPID$_f$ controllers. For FOPID$_f$, the number of function evaluations is slightly higher than GA, but the $f_{AVR}(x)$ is considerably lower. Figures 4.1, 4.2 and 4.3 present the AVR system responses obtained with PIDA, FOPID$_f$, and 2DOFPID$_f$, respectively, and optimized by SA, GA, and ACO-NM.

While results prove that ACO-NM is more efficient than SA and GA, more theoretical developments are needed to understand better how the algorithm works. Some convergence properties for ACO and NM have been presented in [7] and [34], respectively. However, these proofs are not directly valid to the ACO-NM. Therefore, further mathematical development must provide convergence properties such as a performance bound for the algorithm. Moreover, especially for controller tuning, it would be significant to study the interaction between controller parameters further to consider it better during the optimization process and extend this practice to other algorithms.

Table 4.1 Best simulation results obtained with ACO-NM algorithm compared to those presented in Table 3.12

Controller	Algorithm	Bound	K_{acc}	K_{d_1}	K_{p_1}	K_{i_1}	α	β	$f_{AVR}(x)$	nbEvals
PIDA	GA	A	78.88	412.27	1,000	472.70	227.88	946.23	0.6812	9,800
	SA	A	33.1424	342.1740	905.7003	678.3521	65.4891	882.4920	0.6788	10,000
	ACO-NM	A	13.02	263.85	694.41	585.15	18.98	691.94	**0.6741**	**5,155**

Controller	Algorithm	Bound	K_{p_2}	K_{i_2}	K_{d_2}	N_2	λ	ψ	$f_{AVR}(x)$	nbEvals
FOPID$_f$	GA	B	7.8988	292.0336	7.5109	0.1000	480.6571	1.5220	0.8062	**8,550**
	SA	–	–	–	–	–	–	–	–	–
	ACO-NM	B	2.0186	0.5978	0.3261	230.2993	1.0762	0.3249	**0.6657**	9,868

Controller	Algorithm	Bound	K_{p_3}	K_{i_3}	K_{d_3}	N_3	p_1	p_2	$f_{AVR}(x)$	nbEvals
2DOFPID$_f$	GA[a]	A	0.1040	0.1000	0.1000	999.3633	1.7218	2.0342	0.7190	10,000
	SA	B	5.8020	74.3806	1.4227	401.3703	179.9065	412.9024	351.29	10,000
	ACO-NM[a]	B	1.1030	0.5738	0.2924	184.1170	0.9954	0.5994	**0.6972**	7,139

[a]For the same reasons mentioned in Sect. 3.2, the second-best solutions obtained by the algorithms are presented.

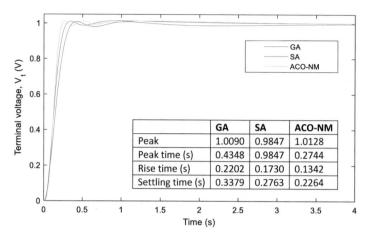

Fig. 4.1 AVR system responses obtained with PIDA optimized by GA, SA, and ACO-NM

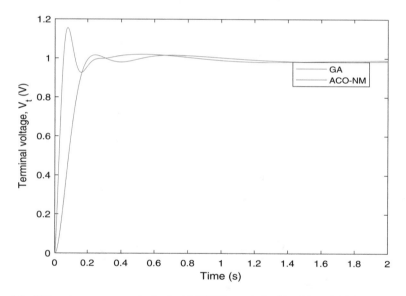

Fig. 4.2 AVR system responses obtained with FOPID$_f$ optimized by GA and ACO-NM

The results substantiate that hybrid algorithms can provide better results than standard algorithms. However, as the number of single algorithms increases continuously, the number of possible hybridizations is growing at a higher speed. Consequently, it is essential to gain theoretical insight into hybridization principles and understand the fundamental reasons and properties of hybridization effectiveness.

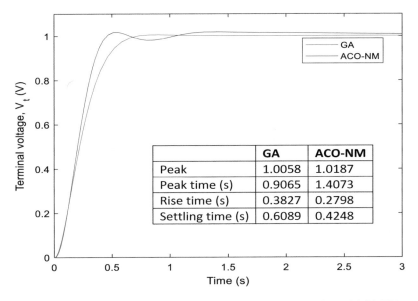

Fig. 4.3 AVR system responses obtained with 2DOFPID$_f$ optimized by GA and ACO-NM

4.4 Conclusion

While the period is empirical assessments were beneficial and allowed to provide solutions to problems where optimization was difficult, the next step for this field is to focus on bringing mathematical proof of algorithm efficiency. This concretely means to explicitly uncover mathematically why the algorithms tend to work so well in practice. Moreover, no free-lunch theorem [35] states that no single algorithm performs best for all kinds of problems. Therefore, it would be a great step forward to establish theoretical insights into which algorithms are more suited to a certain class of problems. More mathematical insight will facilitate to the use of metaheuristics and their hybridization.

References

1. Barr, R.S., Golden, B.L., Kelly, J.P., Resende, M.G., Stewart, W.R.: Designing and reporting on computational experiments with heuristic methods. J. Heurist. **1**(1), 9–32 (1995)
2. Sörensen, K.: Metaheuristics-the metaphor exposed. Int. Trans. Oper. Res. **22**(1), 3–18 (2015)
3. García-Martínez, C., Gutiérrez, P.D., Molina, D., Lozano, M., Herrera, F.: Since CEC 2005 competition on real-parameter optimisation: a decade of research, progress and comparative analysis's weakness. Soft Comput. **21**(19), 5573–5583 (2017)
4. Del Ser, J., et al.: Bio-inspired computation: where we stand and what's next. Swarm Evolut. Comput. **48**, 220–250 (2019)

5. He, J., Yu, X.: Conditions for the convergence of evolutionary algorithms. J. Syst. Archit. **47**(7), 601–612 (2001)
6. He, J., Kang, L.: On the convergence rates of genetic algorithms. Theoret. Comput. Sci. **229**(1–2), 23–39 (1999)
7. Stutzle, T., Dorigo, M.: A short convergence proof for a class of ant colony optimization algorithms. IEEE Trans. Evolut. Comput. **6**(4), 358–365 (2002)
8. Gutjahr, W.J.: ACO algorithms with guaranteed convergence to the optimal solution. Inf. Proc. Lett. **82**(3), 145–153 (2002)
9. Gutjahr, W.J.: A converging ACO algorithm for stochastic combinatorial optimization. In: International Symposium on Stochastic Algorithms, pp. 10–25. Springer, Berlin (2003)
10. Arcuri, A., Fraser, G.: Parameter tuning or default values? An empirical investigation in search-based software engineering. Empir. Softw. Eng. **18**(3), 594–623 (2013)
11. Pedersen, M.E.H.: Good parameters for differential evolution. Magnus Erik Hvass Pedersen, vol. 49 (2010)
12. Blondin, M.J., Sicard, P.: Statistical convergence analysis of ACO-NM for PID controller tuning. In: 2015 IEEE International Conference on Industrial Technology (ICIT), pp. 487–492. IEEE, Piscataway (2015)
13. Blasco Ferragud, F.X.: Control predictivo basado en modelos mediante técnica de optimización heurística. PhD Tesis (en espagnol) Editorial UPV. ISBN 84–699-5429-6 (1999)
14. Affenzeller, M., Wagner, S., Winkler, S.: Self-adaptive population size adjustment for genetic algorithms. In: International Conference on Computer Aided Systems Theory, pp. 820–828. Springer, Berlin (2007)
15. Teo, J.: Exploring dynamic self-adaptive populations in differential evolution. Soft. Comput. **10**, 673–686 (2006). https://doi.org/10.1007/s00500-005-0537-1
16. Hussain, K., Salleh, M.N.M., Cheng, S., Shi, Y.: Metaheuristic research: a comprehensive survey. Artif. Intell. Rev. **52**(4), 2191–2233 (2019)
17. Ting, T.O., Yang, X.S., Cheng, S., Huang, K.: Hybrid Metaheuristic Algorithms: Past, Present, and Future. In: Yang, X.S. (ed.) Recent Advances in Swarm Intelligence and Evolutionary Computation. Studies in Computational Intelligence, vol. 585. Springer, Cham (2015)
18. Blum, C., Puchinger, J., Raidl, G.R., Roli, A.: Hybrid metaheuristics in combinatorial optimization: a survey. Appl. Soft Comput. **11**(6), 4135–4151 (2011)
19. Mafarja, M.M., Mirjalili, S.: Hybrid whale optimization algorithm with simulated annealing for feature selection. Neurocomput. **260**, 302–312 (2017)
20. Wang, X.H., Li, J.J.: Hybrid particle swarm optimization with simulated annealing. In: Proceedings of 2004 International Conference on Machine Learning and Cybernetics (IEEE Cat. No. 04EX826), Vol. 4, pp. 2402–2405. IEEE, Piscataway (2004)
21. Li, W.D., Ong, S.K., Nee, A.Y.: Hybrid genetic algorithm and simulated annealing approach for the optimization of process plans for prismatic parts. Int. J. Prod. Res. **40**(8), 1899–922 (2002)
22. Çelik, E., Öztürk, N.: A hybrid symbiotic organisms search and simulated annealing technique applied to efficient design of PID controller for automatic voltage regulator. Soft. Comput. **22**, 8011–8024 (2018). https://doi.org/10.1007/s00500-018-3432-2
23. Ting, C.J., Chen, C.H.: Combination of multiple ant colony system and simulated annealing for the multidepot vehicle-routing problem with time windows. Transp. Res. Record. **2089**(1), 85–92 (2008)
24. Kim, D.H.: Hybrid GA-BF based intelligent PID controller tuning for AVR system. Appl. Soft Comput. **11**(1), 11–22 (2011)
25. Khadanga, R.K., Satapathy, J.K.: A new hybrid GA-GSA algorithm for tuning damping controller parameters for a unified power flow controller. Int. J. Electr. Power Energy Syst. **73**, 1060–9 (2015)
26. Eslami, M., Shareef, H., Khajehzadeh, M.: Optimal design of damping controllers using a new hybrid artificial bee colony algorithm. Int. J. Electr. Power Energy Syst. **52**, 42–54 (2013)
27. Blondin, M.J., Sanchis, J., Sicard, P., Herrero, J.M.: New optimal controller tuning method for an AVR system using a simplified Ant Colony Optimization with a new constrained Nelder-Mead algorithm. Appl. Soft Comput. **62**, 216–229 (2018)

28. Blondin, M.J., Pardalos, P.M.: A holistic optimization approach for inverted cart-pendulum control tuning. Soft Comput. **24**, 1–17 (2019)
29. Blondin, M.J., Trovão, J.P.: Soft-computing techniques for cruise controller tuning for an off-road electric vehicle. IET Electr. Syst. Transp. **9**(4), 196–205 12 (2019)
30. Mande, D., Blondin, M., Trovão, J.P.F.: Optimization of fractional order PI controller for bidirectional quasi-Z-source inverter used for electric traction system. IET Electrical Systems in Transportation, 2020
31. Blondin, M.J., Sicard, P.: A hybrid ACO and Nelder-Mead constrained algorithm for controller and anti-windup tuning. In: 2014 16th European Conference on Power Electronics and Applications, pp. 1–10. IEEE, Piscataway (2014)
32. Luersen, M.A., Le Riche, R., Guyonm F.: A constrained, globalized, and bounded Nelder-Mead method for engineering optimization. Struct. Multidiscip. Optim. **27**(1–2), 43–54 (2004)
33. Sharma, N., Arun, N., Ravi, V.: An ant colony optimisation and Nelder-Mead simplex hybrid algorithm for training neural networks: an application to bankruptcy prediction in banks. Int. J. Inf. Decision Sci. **5**(2), 188–203 (2013)
34. Lagarias, J.C., Reeds, J.A., Wright, M.H., Wright, P.E.: Convergence properties of the Nelder–Mead simplex method in low dimensions. SIAM J. Optim. **9**(1), 112–47 (1998)
35. Wolpert, D.H., Macready, W.G.: No free lunch theorems for optimization. IEEE Trans. Evolut. Comput. **1**(1), 67–82 (1997)

Printed in the United States
By Bookmasters